Annals of Mathematics Studies

Number 178

The Ambient Metric

Charles Fefferman
C. Robin Graham

PRINCETON UNIVERSITY PRESS
PRINCETON AND OXFORD
2012

Published by Princeton University Press
41 William Street, Princeton, New Jersey 08540

In the United Kingdom: Princeton University Press
6 Oxford Street, Woodstock, Oxfordshire OX20 1TW

Library of Congress Cataloging-in-Publication Data

Fefferman, Charles, 1949-
The ambient metric / Charles Fefferman, C. Robin Graham.
p. cm. – (Annals of mathematics studies ; no.178)
Includes bibliographical references and index.
ISBN 978-0-691-15313-1 (hardback : acid-free paper) – ISBN 978-0-691-15314-8 (pbk. : acid-free paper) 1. Conformal geometry. 2. Conformal invariants. I. Graham, C. Robin, 1954- II. Title.
QA609.F44 2011
516.3′7–dc23 2011023939

British Library Cataloging-in-Publication Data is available

This book has been composed in in LaTeX

The publisher would like to acknowledge the authors of this volume for providing the camera-ready copy from which this book was printed.

Printed on acid-free paper. ∞

press.princeton.edu

Printed in the United States of America

10 9 8 7 6 5 4 3 2 1

To Julie,
with love and admiration

To Charles and Jean Graham,
with gratitude

Contents

The Ambient Metric

Chapter One

Introduction

Conformal geometry is the study of spaces in which one knows how to measure infinitesimal angles but not lengths. A conformal structure on a manifold is an equivalence class of Riemannian metrics, in which two metrics are identified if one is a positive smooth multiple of the other. The study of conformal geometry has a long and venerable history. From the beginning, conformal geometry has played an important role in physical theories.

A striking historical difference between conformal geometry compared with Riemannian geometry is the scarcity of local invariants in the conformal case. Classically known conformally invariant tensors include the Weyl conformal curvature tensor, which plays the role of the Riemann curvature tensor, its three-dimensional analogue the Cotton tensor, and the Bach tensor in dimension four. Further examples are not so easy to come by. By comparison, in the Riemannian case invariant tensors abound. They can be easily constructed by covariant differentiation of the curvature tensor and tensorial operations. The situation is similar for other types of invariant objects, for example for differential operators. Historically, there are scattered examples of conformally invariant operators such as the conformally invariant Laplacian and certain Dirac operators, whereas it is easy to write down Riemannian invariant differential operators, of arbitrary orders and between a wide variety of bundles.

In Riemannian geometry, not only is it easy to write down invariants, it can be shown using Weyl's classical invariant theory for the orthogonal group that all invariants arise by the covariant differentiation and tensorial operations mentioned above. In the case of scalar invariants, this characterization as "Weyl invariants" has had important application in the study of heat asymptotics: one can immediately write down the form of coefficients in the expansion of heat kernels up to the determination of numerical coefficients.

In [FG], we outlined a construction of a nondegenerate Lorentz metric in $n+2$ dimensions associated to an n-dimensional conformal manifold, which we called the ambient metric. This association enables one to construct conformal invariants in n dimensions from pseudo-Riemannian invariants in $n+2$ dimensions, and in particular shows that conformal invariants are plentiful. The construction of conformal invariants is easiest and most effective for scalar invariants: every scalar invariant of metrics in $n+2$ dimensions immediately determines a scalar conformal invariant in n dimensions (which may vanish, however). For other types of invariants, for example for differential operators, some effort is required to derive a conformal invariant from a pseudo-Riemannian invariant in two higher dimensions, but in many cases this can be carried out and has led to important new examples.

The ambient metric is homogeneous with respect to a family of dilations on the $n+2$-dimensional space. It is possible to mod out by these dilations and thereby obtain a metric in $n+1$ dimensions, also associated to the given conformal manifold in n dimensions. This gives the "Poincaré" metric associated to the conformal manifold. The Poincaré metric is complete and the conformal manifold forms its boundary at infinity.

The construction of the ambient and Poincaré metrics associated to a general conformal manifold is motivated by the conformal geometry of the flat model, the sphere S^n, which is naturally described in terms of $n+2$-dimensional Minkowski space. Let

$$Q(x) = \sum_{\alpha=1}^{n+1} (x^\alpha)^2 - \left(x^0\right)^2$$

be the standard Lorentz signature quadratic form on $\mathbb{R}^{n+2}$ and

$$\mathcal{N} = \left\{x \in \mathbb{R}^{n+2} \setminus \{0\} : Q(x) = 0\right\}$$

its null cone. The sphere S^n can be identified with the space of lines in $\mathcal{N}$, with projection $\pi : \mathcal{N} \to S^n$. Let

$$\widetilde{g} = \sum_{\alpha=1}^{n+1} (dx^\alpha)^2 - \left(dx^0\right)^2$$

be the associated Minkowski metric on $\mathbb{R}^{n+2}$. The conformal structure on S^n arises by restriction of $\widetilde{g}$ to $\mathcal{N}$. More specifically, for $x \in \mathcal{N}$ the restriction $\widetilde{g}|_{T_x\mathcal{N}}$ is a degenerate quadratic form because it annihilates the radial vector field $X = \sum_{I=0}^{n+1} x^I \partial_I \in T_x\mathcal{N}$. So $\widetilde{g}|_{T_x\mathcal{N}}$ determines an inner product on $T_x\mathcal{N}/\operatorname{span} X \cong T_{\pi(x)}S^n$. As x varies over a line in $\mathcal{N}$, the resulting inner products on $T_{\pi(x)}S^n$ vary only by scale and are the possible values at $\pi(x)$ for a metric in the conformal class on S^n. The Lorentz group $O(n+1,1)$ acts linearly on $\mathbb{R}^{n+2}$ by isometries of $\widetilde{g}$ preserving $\mathcal{N}$. The induced action on lines in $\mathcal{N}$ therefore preserves the conformal class of metrics and realizes the group of conformal motions of S^n.

If instead of restricting $\widetilde{g}$ to $\mathcal{N}$, we restrict it to the hyperboloid

$$\mathcal{H} = \{x \in \mathbb{R}^{n+2} : Q(x) = -1\},$$

then we obtain the Poincaré metric associated to S^n. Namely, $g_+ := \widetilde{g}|_{T\mathcal{H}}$ is the hyperbolic metric of constant sectional curvature -1. Under an appropriate identification of one sheet of $\mathcal{H}$ with the unit ball in $\mathbb{R}^{n+1}$, g_+ can be realized as the Poincaré metric

$$g_+ = 4\left(1 - |x|^2\right)^{-2} \sum_{\alpha=1}^{n+1} (dx^\alpha)^2,$$

and has the conformal structure on S^n as conformal infinity. The action of $O(n+1,1)$ on $\mathbb{R}^{n+2}$ preserves $\mathcal{H}$. The induced action on $\mathcal{H}$ is clearly by isometries of g_+ and realizes the isometry group of hyperbolic space.

The ambient and Poincaré metrics associated to a general conformal manifold are defined as solutions to certain systems of partial differential equations with initial data determined by the conformal structure. Consider the ambient metric. A conformal class of metrics on a manifold M determines and is determined by its metric bundle $\mathcal{G}$, an $\mathbb{R}_+$-bundle over M. This is the subbundle of symmetric 2-tensors whose sections are the metrics in the conformal class. In the case $M = S^n$, $\mathcal{G}$ can be identified with $\mathcal{N}$ (modulo ± 1). Regard $\mathcal{G}$ as a hypersurface in $\mathcal{G} \times \mathbb{R}$: $\mathcal{G} \cong \mathcal{G} \times \{0\} \subset \mathcal{G} \times \mathbb{R}$. The conditions defining the ambient metric $\widetilde{g}$ are that it be a Lorentz metric defined in a neighborhood of $\mathcal{G}$ in $\mathcal{G} \times \mathbb{R}$ which is homogeneous with respect to the natural dilations on this space, that it satisfy an initial condition on the initial hypersurface $\mathcal{G}$ determined by the conformal structure, and that it be Ricci-flat. The Ricci-flat condition is the system of equations intended to propagate the initial data off of the initial surface.

This initial value problem is singular because the pullback of $\widetilde{g}$ to the initial surface is degenerate. However, for the applications to the construction of conformal invariants, it is sufficient to have formal power series solutions along the initial surface rather than actual solutions in a neighborhood. So we concern ourselves with the formal theory and do not discuss the interesting but more difficult problem of solving the equations exactly.

It turns out that the behavior of solutions of this system depends decisively on the parity of the dimension. When the dimension n of the conformal manifold is odd, there exists a formal power series solution $\widetilde{g}$ which is Ricci-flat to infinite order, and it is unique up to diffeomorphism and up to terms vanishing to infinite order. When $n \geq 4$ is even, there is a solution which is Ricci-flat to order $n/2 - 1$, uniquely determined up to diffeomorphism and up to terms vanishing to higher order. But at this order, the existence of smooth solutions is obstructed by a conformally invariant natural trace-free symmetric 2-tensor, the ambient obstruction tensor. When $n = 4$, the obstruction tensor is the same as the classical Bach tensor. When $n = 2$, there is no obstruction, but uniqueness fails.

It may seem contradictory that for n odd, the solution of a second order initial value problem can be formally determined to infinite order by only one piece of Cauchy data: the initial condition determined by the conformal structure. In fact, there are indeed further formal solutions. These correspond to the freedom of a second piece of initial data at order $n/2$. When n is odd, $n/2$ is half-integral, and this freedom is removed by restricting to formal power series solutions. It is crucially important for the applications to conformal invariants that we are able to uniquely specify an infinite order solution in an invariant way. On the other hand, the additional formal solutions with nontrivial asymptotics at order $n/2$ are also important; they necessarily arise in the global formulation of the existence problem as a boundary value problem at infinity for the Poincaré metric. When n is even, the obstruction to the existence of formal power series solutions can be incorporated into log terms in the expansion, in which case there is again a formally undetermined term at order $n/2$.

The formal theory described above was outlined in [FG], but the details were not given. The first main goal of this monograph is to provide these details. We give the full infinite-order formal theory, including the freedom at order $n/2$ in all dimensions and the precise description of the log terms when $n \geq 4$ is even. This formal theory for the ambient metric forms the content of Chapters 2 and 3. The description of the solutions with freedom at order $n/2$ and log terms extends and sharpens results of Kichenassamy [K]. Convergence of the formal series determined by singular nonlinear initial value problems of this type has been considered by several authors; these results imply that the formal series converge if the data are real-analytic.

In Chapter 4, we define Poincaré metrics: they are formal solutions to the equation $\operatorname{Ric}(g_+) = -ng_+$, and we show how Poincaré metrics are equivalent to ambient metrics satisfying an extra condition which we call straight. Then we use this equivalence to derive the full formal theory for Poincaré metrics from that for ambient metrics. We discuss the "projectively compact" formulation of Poincaré metrics, modeled on the Klein model of hyperbolic space, as well as the usual conformally compact picture. As an application of the formal theory for Poincaré metrics, in Chapter 5 we present a formal power series proof of a result of LeBrun [LeB] asserting the existence and uniqueness of a real-analytic self-dual Einstein metric in dimension 4 defined near the boundary with prescribed real-analytic conformal infinity.

In Chapter 7, we analyze the ambient and Poincaré metrics for locally conformally flat manifolds and for conformal classes containing an Einstein metric. The obstruction tensor vanishes for even dimensional conformal structures of these types. We show that for these special conformal classes, there is a way to uniquely specify the formally undetermined term at order $n/2$ in an invariant way and thereby obtain a unique ambient metric up to terms vanishing to infinite order and up to diffeomorphism, just like in odd dimensions. We derive a formula of Skenderis and Solodukhin [SS] for the ambient or Poincaré metric in the locally conformally flat case which is in normal form relative to an arbitrary metric in the conformal class, and prove a related unique continuation result for hyperbolic metrics in terms of data at conformal infinity. The case $n = 2$ is special for all of these considerations. We also derive the form of the GJMS operators for an Einstein metric.

In [FG], we conjectured that when n is odd, all scalar conformal invariants arise as Weyl invariants constructed from the ambient metric. The second main goal of this monograph is to prove this together with an analogous result when n is even. These results are contained in Theorems 9.2, 9.3, and 9.4. When n is even, we restrict to invariants whose weight w satisfies $-w \leq n$ because of the finite order indeterminacy of the ambient metric: Weyl invariants of higher negative weight may involve derivatives of the ambient metric which are not determined. A particularly interesting phenomenon occurs in dimensions $n \equiv 0 \mod 4$. For all n even, it is the case that all even (i.e., unchanged under orientation reversal) scalar conformal invariants with $-w \leq n$ arise as Weyl invariants of the ambient metric. If $n \equiv 2 \mod 4$, this is also true for odd (i.e., changing sign under orientation reversal) scalar conformal invariants with $-w \leq n$ (in fact, these all vanish). But if $n \equiv 0 \mod 4$, there are odd invariants of weight $-n$ which are exceptional in the sense that they do not arise as Weyl invariants of the ambient metric. The set of such exceptional invariants of weight $-n$ consists precisely of the nonzero elements of the vector space spanned by the Pontrjagin invariants whose integrals give the Pontrjagin numbers of a compact oriented n-dimensional manifold (Theorem 9.3).

The parabolic invariant theory needed to prove these results was developed in [BEGr], including the observation of the existence of exceptional invariants. But substantial work is required to reduce the theorems in Chapter 9 to the results of [BEGr]. To understand this, we briefly review how Weyl's characterization of scalar Riemannian invariants is proved.

Recall that Weyl's theorem for even invariants states that every even scalar Riemannian invariant is a linear combination of complete contractions of the form $\operatorname{contr}(\nabla^{r_1} R \otimes \cdots \otimes \nabla^{r_L} R)$, where the r_i are nonnegative integers, $\nabla^r R$ denotes the r-th covariant derivative of the curvature tensor, and contr denotes a metric contraction with respect to some pairing of all the indices. There are two main steps in the proof of Weyl's theorem. The first is to show that any scalar Riemannian invariant can be written as a polynomial in the components of the covariant derivatives of the curvature tensor which is invariant under the orthogonal group $O(n)$. Since a Riemannian invariant by definition is a polynomial in the Taylor coefficients of the metric in local coordinates whose value is independent of the choice of coordinates, one must pass from Taylor coefficients of the metric to covariant derivatives of curvature. This passage is carried out using geodesic normal coordinates. We refer to the result stating that the map from Taylor coefficients of the metric in geodesic normal coordinates to covariant derivatives of curvature is an $O(n)$-equivariant isomorphism as the jet isomorphism theorem for Riemannian geometry. Once the jet isomorphism theorem has been established, one is left with the algebraic problem of identifying the $O(n)$-invariant polynomials in the covariant derivatives of curvature. This is solved by Weyl's classical invariant theory for the orthogonal group.

In the conformal case, the role of the covariant derivatives of the curvature tensor is played by the covariant derivatives of the curvature tensor of the ambient metric. These tensors are of course defined on the ambient space. But when evaluated on the initial surface, their components relative to a suitable frame determined by a choice of metric in the conformal class define tensors on the base conformal manifold, which we call conformal curvature tensors. For example, the conformal curvature tensors defined by the curvature tensor of the ambient metric itself (i.e., with no ambient covariant derivatives) are the classical Weyl, Cotton, and Bach tensors (except that in dimension 4, the Bach tensor does not arise as a conformal curvature tensor because of the indeterminacy of the ambient metric). The covariant derivatives of curvature of the ambient metric satisfy identities and relations beyond those satisfied for general metrics owing to its homogeneity and Ricci-flatness. We derive these identities in Chapter 6. We also derive the transformation laws for the conformal curvature tensors under conformal change. Of all the conformal curvature tensors, only the Weyl tensor (and Cotton tensor in dimension 3) are conformally invariant. The transformation law of any other conformal curvature tensor involves only first derivatives of the conformal factor and "earlier" conformal curvature tensors. These transformation laws may also be interpreted in terms of tractors. When n is even, the definitions of the conformal curvature tensors and the identities which they satisfy are restricted by the finite order indeterminacy of the ambient metric. The ambient obstruction tensor is not a conformal curvature tensor; it lies at the boundary of the range for which they are defined. But it may be regarded as the residue of an analytic continuation in the dimension of conformal curvature tensors in higher dimensions (Proposition 6.7).

Having understood the properties of the conformal curvature tensors, the next step in the reduction of the theorems in Chapter 9 to the results of [BEGr] is to formulate and prove a jet isomorphism theorem for conformal geometry, in order to know that a scalar conformal invariant can be written in terms of conformal curvature tensors. The Taylor expansion of the metric on the base manifold in geodesic normal coordinates can be further simplified since one now has the freedom to change the metric by a conformal factor as well as by a diffeomorphism. This leads to a "conformal normal form" in which part of the base curvature is normalized away to all orders. Then the conformal jet isomorphism theorem states that the map from the Taylor coefficients of a metric in conformal normal form to the space of all conformal curvature tensors, realized as covariant derivatives of ambient curvature, is an isomorphism. Again, the spaces must be truncated at finite order in even dimensions. The proof of the conformal jet isomorphism theorem is much more involved than in the Riemannian case; it is necessary to relate the normalization conditions in the conformal normal form to the precise identities and relations satisfied by the ambient covariant derivatives of curvature. We carry this out in Chapter 8 by making a direct algebraic study of these relations and of the map from jets of normalized metrics to conformal curvature tensors. A more conceptual proof of the conformal jet isomorphism theorem due to the second author and K. Hirachi uses an ambient lift of the conformal deformation complex and is outlined in [Gr3].

The orthogonal group plays a central role in Riemannian geometry because it is the isotropy group of a point in the group of isometries of the flat model $\mathbb{R}^n$. The analogous group for conformal geometry is the isotropy subgroup $P \subset O(n+1,1)$ of the conformal group fixing a point in S^n, i.e., a null line. Because of its algebraic structure, P is referred to as a parabolic subgroup of $O(n+1,1)$. Just as geodesic normal coordinates are determined up to the action of $O(n)$ in the Riemannian case, the equivalent conformal normal forms for a metric at a given point are determined up to an action of P. Since P is a matrix group in $n+2$ dimensions, there is a natural tensorial action of P on the space of covariant derivatives of ambient curvature, and the conformal transformation law for conformal curvature tensors established in Chapter 6 implies that the map from jets of metrics in conformal normal form to conformal curvature tensors is P-equivariant.

The jet isomorphism theorem reduces the study of conformal invariants to the purely algebraic matter of understanding the P-invariants of the space of covariant derivatives of ambient curvature. This space is nonlinear since the Ricci identity for commuting covariant derivatives is nonlinear in curvature and its derivatives. The results of [BEGr] identify the P-invariants of the linearization of this space. So the last steps, carried out in Chapter 9, are to formulate the results about scalar invariants, to use the jet isomorphism theorem to reduce these results to algebraic statements in invariant theory for P, and finally to reduce the invariant theory for the actual nonlinear space to that for its linearization. The treatment in Chapters 8 and 9 is inspired by, and to some degree follows, the treatment in [F] in the case of CR geometry.

Our work raises the obvious question of extending the theory to higher orders in even dimensions. This has recently been carried out by the second author and K. Hirachi. An extension to all orders of the ambient metric construction, jet isomorphism theorem, and invariant theory has been announced in [GrH2], [Gr3] inspired by the work of Hirachi [Hi] in the CR case. The log terms in the expansion of an ambient metric are modified by taking the log of a defining function homogeneous of degree 2 rather than homogeneous of degree 0. This makes it possible to define the smooth part of an ambient metric with log terms in an invariant way. The smooth part is smooth and homogeneous but no longer Ricci flat to infinite order. There is a family of such smooth parts corresponding to different choices of the ambiguity at order $n/2$. They can be used to formulate a jet isomorphism theorem and to construct invariants, and the main conclusion is that up to a linear combination of finitely many exceptional odd invariants in dimensions $n \equiv 0 \mod 4$ which can be explicitly identified, all scalar conformal invariants arise from the ambient metric. An alternate development of a conformal invariant theory based on tractor calculus is given in general dimensions in [Go1].

A sizeable literature concerning the ambient and Poincaré metrics has arisen since the publication of [FG]. The subject has been greatly stimulated by its relevance in the study of the AdS/CFT correspondence in physics. We have tried to indicate some of the most relevant references of which we are aware without attempting to be exhaustive. Juhl's recent book [J] has some overlap with our material and much more in the direction of Q-curvature and holography.

A construction equivalent to the ambient metric was derived by Haantjes and Schouten in [HS]. They obtained a version of the expansion for straight ambient metrics, to infinite order in odd dimensions and up to the obstruction in even dimensions. In particular, they showed that there is an obstruction in even dimensions $n \geq 4$ and calculated that it is the Bach tensor in dimension 4. They observed that the obstruction vanishes for conformally Einstein metrics and in this case derived the conformally invariant normalization uniquely specifying an infinite order ambient metric in even dimensions. They also obtained the infinite-order expansion in the case of dimension 2, including the precise description of the non-uniqueness of solutions. Haantjes and Schouten did not consider applications to conformal invariants and, unfortunately, it seems that their work was largely forgotten.

We are grateful to the National Science Foundation for support. In particular, the second author was partially supported by NSF grants # DMS 0505701 and 0906035 during the preparation of this manuscript.

Throughout, by smooth we will mean infinitely differentiable. Manifolds are assumed to be smooth and second countable; hence paracompact. Our setting is primarily algebraic, so we work with metrics of general signature. In tensorial expressions, we denote by parentheses (ijk) symmetrization and by brackets $[ijk]$ skew-symmetrization over the enclosed indices.

Chapter Two

Ambient Metrics

Let M be a smooth manifold of dimension $n \geq 2$ equipped with a conformal class $[g]$. Here, g is a smooth pseudo-Riemannian metric of signature (p, q) on M and $[g]$ consists of all metrics

$$\widehat{g} = e^{2\Upsilon} g$$

on M, where Υ is any smooth real-valued function on M.

The space $\mathcal{G}$ consists of all pairs (h, x), where $x \in M$, and h is a symmetric bilinear form on T_xM satisfying $h = s^2 g_x$ for some $s \in \mathbb{R}_+$. Here and below, g_x denotes the symmetric bilinear form on T_xM induced by the metric g. We write $\pi : \mathcal{G} \to M$ for the projection map $(h, x) \mapsto x$. Also, for $s \in \mathbb{R}_+$, we define the "dilation" $\delta_s : \mathcal{G} \to \mathcal{G}$ by setting $\delta_s(h, x) = (s^2 h, x)$. The space $\mathcal{G}$, equipped with the projection π and the dilations $(\delta_s)_{s \in \mathbb{R}_+}$, is an $\mathbb{R}_+$-bundle. We call it the *metric bundle* for $(M, [g])$. We denote by $T = \frac{d}{ds}\delta_s|_{s=1}$ the vector field on $\mathcal{G}$ which is the infinitesimal generator of the dilations δ_s.

There is a tautological symmetric 2-tensor $\mathbf{g}_0$ on $\mathcal{G}$, defined as follows. Let $z = (h, x) \in \mathcal{G}$, and let $\pi_* : T\mathcal{G} \to TM$ be the differential of the map π. Then, for tangent vectors $X, Y \in T_z\mathcal{G}$, we define $\mathbf{g}_0(X, Y) = h(\pi_* X, \pi_* Y)$. The 2-tensor $\mathbf{g}_0$ is homogeneous of degree 2 with respect to the dilations δ_s. That is, $\delta_s^* \mathbf{g}_0 = s^2 \mathbf{g}_0$. One checks easily that the $\mathbb{R}_+$-bundle $\mathcal{G}$, the maps δ_s and π, and the tautological 2-tensor $\mathbf{g}_0$ on $\mathcal{G}$, all depend only on the conformal class $[g]$, and are independent of the choice of the representative g. However, once we fix a representative g, we obtain a trivialization of the bundle $\mathcal{G}$. In fact, we identify

$$(t, x) \in \mathbb{R}_+ \times M \quad \text{with} \quad (t^2 g_x, x) \in \mathcal{G}\,.$$

In terms of this identification, the dilations δ_s, the projection π, the vector field T, and the tautological 2-tensor $\mathbf{g}_0$ are given by

$$\delta_s : (t, x) \mapsto (st, x), \qquad \pi : (t, x) \mapsto x, \qquad T = t\partial_t, \qquad \mathbf{g}_0 = t^2 \pi^* g.$$

The metric g can be regarded as a section of the bundle $\mathcal{G}$. The image of this section is the submanifold of $\mathcal{G}$ given by $t = 1$. The choice of g also determines a horizontal subspace $\mathcal{H}_z \subset T_z\mathcal{G}$ for each $z \in \mathcal{G}$, namely $\mathcal{H}_z = \ker(dt)_z$. In terms of another representative $\widehat{g} = e^{2\Upsilon}g$ of the same conformal class, we obtain another trivialization of $\mathcal{G}$, by identifying

$$(\widehat{t}, x) \in \mathbb{R}_+ \times M \quad \text{with} \quad (\widehat{t}^2 \widehat{g}_x, x) \in \mathcal{G}.$$

The two trivializations are then related by the formula

$$\widehat{t} = e^{-\Upsilon(x)} t \text{ for } \ (t, x) \in \mathbb{R}_+ \times M. \tag{2.1}$$

If $(x^1, \cdots, x^n)$ are local coordinates on an open set U in M, and if g is given in these coordinates as $g = g_{ij}(x)dx^i dx^j$, then $(t, x^1, \cdots, x^n)$ are local coordinates on $\pi^{-1}(U)$, and $\mathbf{g}_0$ is given by

$$\mathbf{g}_0 = t^2 g_{ij}(x) dx^i dx^j.$$

The horizontal subspace $\mathcal{H}_z$ is the span of $\{\partial_{x^1}, \cdots, \partial_{x^n}\}$.

Consider now the space $\mathcal{G} \times \mathbb{R}$. We write points of $\mathcal{G} \times \mathbb{R}$ as (z, ρ), with $z \in \mathcal{G}$, $\rho \in \mathbb{R}$. The dilations δ_s extend to $\mathcal{G} \times \mathbb{R}$ acting in the first factor alone, and we denote these dilations also by δ_s. The infinitesimal dilation T also extends to $\mathcal{G} \times \mathbb{R}$. We embed $\mathcal{G}$ into $\mathcal{G} \times \mathbb{R}$ by $\iota : z \mapsto (z, 0)$ for $z \in \mathcal{G}$. Note that ι commutes with dilations. If g is a representative for the conformal structure with associated fiber coordinate t, and if $(x^1, \cdots, x^n)$ are local coordinates on M as above, then $(t, x^1, \cdots, x^n, \rho)$ are local coordinates on $\mathcal{G} \times \mathbb{R}$. We use 0 to label the t-component, ∞ to label the ρ-component, lowercase Latin letters for M, and capital Latin letters for $\mathcal{G} \times \mathbb{R}$. Even without a choice of coordinates on M, we can use $0, i, \infty$ as labels for the components relative to the identification $\mathcal{G} \times \mathbb{R} \simeq \mathbb{R}_+ \times M \times \mathbb{R}$ induced by the choice of representative metric g. Such an interpretation is coordinate-free and global on M.

Definition 2.1. A *pre-ambient space* for $(M, [g])$, where $[g]$ is a conformal class of signature (p, q) on M, is a pair $(\widetilde{\mathcal{G}}, \widetilde{g})$, where

(1) $\widetilde{\mathcal{G}}$ is a dilation-invariant open neighborhood of $\mathcal{G} \times \{0\}$ in $\mathcal{G} \times \mathbb{R}$;

(2) $\widetilde{g}$ is a smooth metric of signature $(p+1, q+1)$ on $\widetilde{\mathcal{G}}$;

(3) $\widetilde{g}$ is homogeneous of degree 2 on $\widetilde{\mathcal{G}}$ (i.e., $\delta_s^* \widetilde{g} = s^2 \widetilde{g}$, for $s \in \mathbb{R}_+$);

(4) The pullback $\iota^* \widetilde{g}$ is the tautological tensor $\mathbf{g}_0$ on $\mathcal{G}$.

If $(\widetilde{\mathcal{G}}, \widetilde{g})$ is a pre-ambient space, the metric $\widetilde{g}$ is called a *pre-ambient metric.* If the dimension n of M is odd or $n = 2$, then a pre-ambient space $(\widetilde{\mathcal{G}}, \widetilde{g})$ is called an *ambient space* for $(M, [g])$ provided we have

(5) $\operatorname{Ric}(\widetilde{g})$ vanishes to infinite order at every point of $\mathcal{G} \times \{0\}$.

We prepare to define ambient spaces in the even-dimensional case. Let S_{IJ} be a smooth symmetric 2-tensor field on an open neighborhood of $\mathcal{G} \times \{0\}$ in $\mathcal{G} \times \mathbb{R}$. For an integer $m \geq 0$, we write $S_{IJ} = O^+_{IJ}(\rho^m)$ if

(i) $S_{IJ} = O(\rho^m)$; and

(ii) For each point $z \in \mathcal{G}$, the symmetric 2-tensor $(\iota^*(\rho^{-m}S))(z)$ is of the form π^*s for some symmetric 2-tensor s at $x = \pi(z) \in M$ satisfying $\mathrm{tr}_{g_x} s = 0$. The symmetric 2-tensor s is allowed to depend on z, not just on x.

In terms of components relative to a choice of representative metric g, $S_{IJ} = O^+_{IJ}(\rho^m)$ if and only if all components satisfy $S_{IJ} = O(\rho^m)$ and if in addition one has that S_{00}, S_{0i} and $g^{ij}S_{ij}$ are $O(\rho^{m+1})$. The condition $S_{IJ} = O^+_{IJ}(\rho^m)$ is easily seen to be preserved by diffeomorphisms ϕ on a neighborhood of $\mathcal{G} \times \{0\}$ in $\mathcal{G} \times \mathbb{R}$ satisfying $\phi|_{\mathcal{G}\times\{0\}} = \text{identity}$.

Now suppose $(\widetilde{\mathcal{G}}, \widetilde{g})$ is a pre-ambient space for $(M, [g])$, with $n = \dim M$ even and $n \geq 4$. We say that $(\widetilde{\mathcal{G}}, \widetilde{g})$ is an *ambient space* for $(M, [g])$, provided we have

(5′) $\mathrm{Ric}(\widetilde{g}) = O^+_{IJ}(\rho^{n/2-1})$.

If $(\widetilde{\mathcal{G}}, \widetilde{g})$ is an ambient space, the metric $\widetilde{g}$ is called an *ambient metric*.

Next, we define a notion of ambient equivalence for pre-ambient spaces.

Definition 2.2. Let $(\widetilde{\mathcal{G}}_1, \widetilde{g}_1)$ and $(\widetilde{\mathcal{G}}_2, \widetilde{g}_2)$ be two pre-ambient spaces for $(M, [g])$. We say that $(\widetilde{\mathcal{G}}_1, \widetilde{g}_1)$ and $(\widetilde{\mathcal{G}}_2, \widetilde{g}_2)$ are *ambient-equivalent* if there exist open sets $\mathcal{U}_1 \subset \widetilde{\mathcal{G}}_1$, $\mathcal{U}_2 \subset \widetilde{\mathcal{G}}_2$ and a diffeomorphism $\phi : \mathcal{U}_1 \to \mathcal{U}_2$, with the following properties:

(1) $\mathcal{U}_1$ and $\mathcal{U}_2$ both contain $\mathcal{G} \times \{0\}$;

(2) $\mathcal{U}_1$ and $\mathcal{U}_2$ are dilation-invariant and ϕ commutes with dilations;

(3) The restriction of ϕ to $\mathcal{G} \times \{0\}$ is the identity map;

(4) If $n = \dim M$ is odd, then $\widetilde{g}_1 - \phi^*\widetilde{g}_2$ vanishes to infinite order at every point of $\mathcal{G} \times \{0\}$.

(4′) If $n = \dim M$ is even, then $\widetilde{g}_1 - \phi^*\widetilde{g}_2 = O^+_{IJ}(\rho^{n/2})$.

It is easily seen that ambient-equivalence is an equivalence relation.

One of the main results of this monograph is the following.

Theorem 2.3. *Let $(M, [g])$ be a smooth manifold of dimension $n \geq 2$, equipped with a conformal class. Then there exists an ambient space for $(M, [g])$. Also, any two ambient spaces for $(M, [g])$ are ambient-equivalent.*

It is clear when n is odd that a pre-ambient space is an ambient space provided it is ambient-equivalent to an ambient space. The kinds of arguments we use in Chapter 3 can be used to show that this is also true if n is even and $n \geq 4$. Thus for $n > 2$, an ambient space for $(M, [g])$ is determined precisely up to ambient-equivalence. This is clearly not true when $n = 2$: changing the metric at high finite order generally affects the infinite-order vanishing of the Ricci curvature. The uniqueness of ambient metrics when $n = 2$ will be clarified in Theorem 3.7.

Our proof of Theorem 2.3 will establish an additional important property of ambient metrics. In Chapter 3 we will prove the following two propositions.

Proposition 2.4. *Let $(\widetilde{\mathcal{G}}, \widetilde{g})$ be a pre-ambient space for $(M, [g])$. There is a dilation-invariant open set $\mathcal{U} \subset \widetilde{\mathcal{G}}$ containing $\mathcal{G} \times \{0\}$ such that the following three conditions are equivalent.*

(1) $\widetilde{\nabla} T = Id$ *on* $\mathcal{U}$.

(2) $2T \lrcorner \widetilde{g} = d(\|T\|^2)$ *on* $\mathcal{U}$.

(3) For each $p \in \mathcal{U}$, the parametrized dilation orbit $s \mapsto \delta_s p$ is a geodesic for $\widetilde{g}$.

In (1), $\widetilde{\nabla}$ denotes the covariant derivative with respect to the Levi-Civita connection of $\widetilde{g}$. So $\widetilde{\nabla} T$ is a $(1,1)$-tensor on $\mathcal{U}$, and the requirement is that it be the identity endomorphism at each point. In (2), $\|T\|^2 = \widetilde{g}(T,T)$.

Definition 2.5. A pre-ambient space $(\widetilde{\mathcal{G}}, \widetilde{g})$ for $(M, [g])$ will be said to be *straight* if the equivalent properties of Proposition 2.4 hold with $\mathcal{U} = \widetilde{\mathcal{G}}$. In this case, the pre-ambient metric $\widetilde{g}$ is also said to be straight.

Note that if $(\widetilde{\mathcal{G}}, \widetilde{g})$ is a straight pre-ambient space for $(M, [g])$ and ϕ is a diffeomorphism of a dilation-invariant open neighborhood $\mathcal{U}$ of $\mathcal{G} \times \{0\}$ into $\widetilde{\mathcal{G}}$ which commutes with dilations and satisfies that $\phi|_{\mathcal{G} \times \{0\}}$ is the identity map, then the pre-ambient space $(\mathcal{U}, \phi^* \widetilde{g})$ is also straight.

Proposition 2.6. *Let $(M, [g])$ be a smooth manifold of dimension $n > 2$ equipped with a conformal class. Then there exists a straight ambient space for $(M, [g])$. Moreover, if $\widetilde{g}$ is any ambient metric for $(M, [g])$, there is a straight ambient metric $\widetilde{g}'$ such that if n is odd, then $\widetilde{g} - \widetilde{g}'$ vanishes to infinite order at $\mathcal{G} \times \{0\}$, while if n is even, then $\widetilde{g} - \widetilde{g}' = O^+_{IJ}(\rho^{n/2})$.*

Because of Proposition 2.6, one can usually restrict attention to straight ambient spaces. Observe that the second statement of Proposition 2.6 follows from the first statement, the uniqueness up to ambient-equivalence in Theorem 2.3, and the diffeomorphism-invariance of the straightness condition.

In this chapter, we will begin the proof of Theorem 2.3 by using a diffeomorphism to bring a pre-ambient metric into a normal form relative to a choice of representative of the conformal class. Chapter 3 will analyze metrics in that normal form and complete the proof of Theorem 2.3. We start by formulating the normal form condition.

Definition 2.7. A pre-ambient space $(\widetilde{\mathcal{G}}, \widetilde{g})$ for $(M, [g])$ is said to be in *normal form* relative to a representative metric g if the following three conditions hold:

(1) For each fixed $z \in \mathcal{G}$, the set of all $\rho \in \mathbb{R}$ such that $(z, \rho) \in \widetilde{\mathcal{G}}$ is an open interval I_z containing 0.

(2) For each $z \in \mathcal{G}$, the parametrized curve $I_z \ni \rho \mapsto (z, \rho)$ is a geodesic for the metric $\widetilde{g}$.

(3) Let us write (t, x, ρ) for a point of $\mathbb{R}_+ \times M \times \mathbb{R} \simeq \mathcal{G} \times \mathbb{R}$ under the identification induced by g, as discussed above. Then, at each point $(t, x, 0) \in \mathcal{G} \times \{0\}$, the metric tensor $\widetilde{g}$ takes the form

$$\widetilde{g} = \mathbf{g}_0 + 2t\, dt\, d\rho\,. \tag{2.2}$$

The main result proved in this chapter is the following.

Proposition 2.8. *Let $(M, [g])$ be a smooth manifold equipped with a conformal class, let g be a representative of the conformal class, and let $(\widetilde{\mathcal{G}}, \widetilde{g})$ be a pre-ambient space for $(M, [g])$. Then there exists a dilation-invariant open set $\mathcal{U} \subset \mathcal{G} \times \mathbb{R}$ containing $\mathcal{G} \times \{0\}$ on which there is a unique diffeomorphism ϕ from $\mathcal{U}$ into $\widetilde{\mathcal{G}}$, such that ϕ commutes with dilations, $\phi|_{\mathcal{G}\times\{0\}}$ is the identity map, and such that the pre-ambient space $(\mathcal{U}, \phi^*\widetilde{g})$ is in normal form relative to g.*

Thus, once we have picked a representative g of the conformal class $[g]$, we can uniquely place any given pre-ambient metric into normal form by a diffeomorphism ϕ. In Proposition 2.8, note that $(\mathcal{U}, \phi^*\widetilde{g})$ is ambient-equivalent to $(\widetilde{\mathcal{G}}, \widetilde{g})$.

In Chapter 3, we will establish the following result.

Theorem 2.9. *Let M be a smooth manifold of dimension $n \geq 2$ and g a smooth metric on M.*

(A) There exists an ambient space $(\widetilde{\mathcal{G}}, \widetilde{g})$ for $(M, [g])$ which is in normal form relative to g.

(B) Suppose that $(\widetilde{\mathcal{G}}_1, \widetilde{g}_1)$ and $(\widetilde{\mathcal{G}}_2, \widetilde{g}_2)$ are two ambient spaces for $(M, [g])$, both of which are in normal form relative to g. If n is odd, then $\widetilde{g}_1 - \widetilde{g}_2$ vanishes to infinite order at every point of $\mathcal{G} \times \{0\}$. If n is even, then $\widetilde{g}_1 - \widetilde{g}_2 = O^+_{IJ}(\rho^{n/2})$.

Proof of Theorem 2.3 using Proposition 2.8 and Theorem 2.9. Given $(M,[g])$, we pick a representative g and invoke Theorem 2.9(A). Thus, there exists an ambient space for $(M,[g])$. For the uniqueness, let $(\widetilde{\mathcal{G}}_1,\widetilde{g}_1)$ and $(\widetilde{\mathcal{G}}_2,\widetilde{g}_2)$ be ambient spaces for $(M,[g])$. Again, we pick a representative g. Applying Proposition 2.8, we find that $(\widetilde{\mathcal{G}}_1,\widetilde{g}_1)$ is ambient-equivalent to an ambient space in normal form relative to g. Similarly, $(\widetilde{\mathcal{G}}_2,\widetilde{g}_2)$ is ambient-equivalent to an ambient space also in normal form relative to g. Theorem 2.9(B) shows that these two ambient spaces in normal form are ambient-equivalent. Consequently, $(\widetilde{\mathcal{G}}_1,\widetilde{g}_1)$ is ambient-equivalent to $(\widetilde{\mathcal{G}}_2,\widetilde{g}_2)$. □

The rest of this chapter is devoted to the proof of Proposition 2.8. We first formulate a notion which will play a key role in the proof. Let $(M,[g])$ be a conformal manifold, let $(\widetilde{\mathcal{G}},\widetilde{g})$ be a pre-ambient space for $(M,[g])$, and let g be a metric in the conformal class. Recall that g determines the fiber coordinate $t:\mathcal{G}\to\mathbb{R}_+$ and the horizontal subbundle $\mathcal{H}=\ker(dt)\subset T\mathcal{G}$. For $z\in\mathcal{G}$, we may view $\mathcal{H}_z$ as a subspace of $T_{(z,0)}(\mathcal{G}\times\mathbb{R})$ via the inclusion $\iota:\mathcal{G}\to\mathcal{G}\times\mathbb{R}$. For $z\in\mathcal{G}$, we say that a vector $V\in T_{(z,0)}(\mathcal{G}\times\mathbb{R})$ is a g-transversal for $\widetilde{g}$ at $(z,0)$ if it satisfies the following conditions:

$$\begin{aligned}&\widetilde{g}(V,T)=t^2\\&\widetilde{g}(V,X)=0 \text{ for all } X\in\mathcal{H}_z\\&\widetilde{g}(V,V)=0.\end{aligned}\tag{2.3}$$

In the first line, t denotes the fiber coordinate for the point z. The motivation for this definition is the observation that $V=\partial_\rho$ is a g-transversal for $\widetilde{g}$ if $\widetilde{g}$ satisfies condition (3) of Definition 2.7. In general we have the following elementary result.

Lemma 2.10. *For each $z\in\mathcal{G}$, there exists one and only one g-transversal V_z for $\widetilde{g}$ at $(z,0)$. Moreover, V_z is transverse to $\mathcal{G}\times\{0\}$, V_z depends smoothly on z, and V_z is dilation-invariant in the sense that $(\delta_s)_*V_z=V_{\delta_s(z)}$ for $s\in\mathbb{R}_+$ and $z\in\mathcal{G}$.*

Proof. Recall the identification $\mathcal{G}\times\mathbb{R}\simeq\mathbb{R}_+\times M\times\mathbb{R}$ determined by g. In terms of the coordinates (t,x^i,ρ) induced by a choice of local coordinates on M, we can express $V=V^0\partial_t+V^i\partial_{x^i}+V^\infty\partial_\rho$. We can also express $\widetilde{g}$ in these coordinates. By condition (4) of Definition 2.1, at a point $(t,x,0)$, $\widetilde{g}$ takes the form

$$\widetilde{g}=t^2g_{ij}(x)dx^idx^j+2\widetilde{g}_{0\infty}dtd\rho+2\widetilde{g}_{j\infty}dx^jd\rho+\widetilde{g}_{\infty\infty}(d\rho)^2$$

where $\widetilde{g}_{0\infty}$, $\widetilde{g}_{j\infty}$ and $\widetilde{g}_{\infty\infty}$ depend on (t,x). Nondegeneracy of $\widetilde{g}$ implies that $\widetilde{g}_{0\infty}\neq 0$. The conditions (2.3) defining a g-transversal become

$$\widetilde{g}_{0\infty}V^\infty=t,\qquad t^2g_{ij}V^i+\widetilde{g}_{j\infty}V^\infty=0,$$

$$2\widetilde{g}_{0\infty}V^0V^\infty + t^2g_{ij}V^iV^j + 2\widetilde{g}_{i\infty}V^iV^\infty + \widetilde{g}_{\infty\infty}(V^\infty)^2 = 0.$$

It is clear that these equations can be successively solved uniquely for V^∞, V^i; V^0, and the other conclusions of Lemma 2.10 follow easily from the smoothness and homogeneity properties defining a pre-ambient space. □

Proof of Proposition 2.8. For $z \in \mathcal{G}$, let V_z be the g-transversal for $\widetilde{g}$ at $(z,0)$ given by Lemma 2.10. Let $\lambda \mapsto \phi(z,\lambda) \in \widetilde{\mathcal{G}}$ be a (parametrized) geodesic for $\widetilde{g}$, with initial conditions

$$\phi(z,0) = (z,0) \qquad \partial_\lambda\phi(z,\lambda)|_{\lambda=0} = V_z\,. \tag{2.4}$$

Since $\widetilde{g}$ needn't be geodesically complete, $\phi(z,\lambda)$ is defined only for (z,λ) in an open neighborhood $\mathcal{U}_0$ of $\mathcal{G}\times\{0\}$ in $\mathcal{G}\times\mathbb{R}$. Since $\widetilde{g}$ and V_z are homogeneous with respect to the dilations, we may take $\mathcal{U}_0$ to be dilation-invariant. Thus, $\phi : \mathcal{U}_0 \to \widetilde{\mathcal{G}}$ is a smooth map, commuting with dilations, and satisfying (2.4).

Since V_z is transverse to $\mathcal{G}\times\{0\}$, it follows that $\mathcal{U}_1 = \{(z,\lambda) \in \mathcal{U}_0 : \det\phi'(z,\lambda) \neq 0\}$ is a dilation-invariant open neighborhood of $\mathcal{G}\times\{0\}$ in $\mathcal{U}_0$. Thus, ϕ is a local diffeomorphism from $\mathcal{U}_1$ into $\widetilde{\mathcal{G}}$, commuting with dilations. Moreover, by definition of ϕ, we have

> Let $z \in \mathcal{G}$ and let I be an interval containing 0. Assume that $(z,\lambda) \in \mathcal{U}_1$ for all $\lambda \in I$. Then $I \ni \lambda \mapsto \phi(z,\lambda)$ is a geodesic for $\widetilde{g}$, with initial conditions (2.4).

The map ϕ need not be globally one-to-one on $\mathcal{U}_1$. However, arguing as in the proof of the Tubular Neighborhood Theorem (see, e.g., [L]), one concludes that there exists a dilation-invariant open neighborhood $\mathcal{U}_2$ of $\mathcal{G}\times\{0\}$ in $\mathcal{U}_1$, such that $\phi|_{\mathcal{U}_2}$ is globally one-to-one. Thus, ϕ is a diffeomorphism from $\mathcal{U}_2$ to a dilation-invariant open subset of $\widetilde{\mathcal{G}}$ containing $\mathcal{G}\times\{0\}$.

Next, we define $\mathcal{U} = \{(z,\lambda) \in \mathcal{U}_2 : (z,\mu) \in \mathcal{U}_2$ for all $\mu \in \mathbb{R}$ for which $|\mu| \leq |\lambda|\}$. Thus, $\mathcal{U}$ is a dilation-invariant open neighborhood of $\mathcal{G}\times\{0\}$ in $\mathcal{U}_2$. Moreover, for each fixed $z \in \mathcal{G}$, $\{\lambda \in \mathbb{R} : (z,\lambda) \in \mathcal{U}\}$ is an open interval I_z containing 0. It follows that for each fixed $z \in \mathcal{G}$, the parametrized curve $I_z \ni \lambda \mapsto \phi(z,\lambda)$ is a geodesic for the metric $\widetilde{g}$.

Since $(\widetilde{\mathcal{G}},\widetilde{g})$ is a pre-ambient space for $(M,[g])$, so is $(\mathcal{U},\phi^*\widetilde{g})$. For each fixed $z \in \mathcal{G}$, the parametrized curve $I_z \ni \lambda \mapsto (z,\lambda)$ is a geodesic for $\phi^*\widetilde{g}$. From the facts that V satisfies (2.3) and ϕ satisfies (2.4), it follows that under the identification $\mathbb{R}_+\times M\times\mathbb{R} \simeq \mathcal{G}\times\mathbb{R}$ induced by g, we have at $\lambda = 0$

$$(\phi^*\widetilde{g})(\partial_\lambda, T) = t^2,$$
$$(\phi^*\widetilde{g})(\partial_\lambda, X) = 0 \text{ for } X \in TM,$$
$$(\phi^*\widetilde{g})(\partial_\lambda, \partial_\lambda) = 0.$$

Together with property (4) of Definition 2.1 of the pre-ambient space $(\widetilde{\mathcal{G}},\widetilde{g})$, these equations show that $\phi^*\widetilde{g} = \mathbf{g}_0 + 2t\,dt\,d\lambda$ when $\lambda = 0$. This establishes the existence part of Proposition 2.8. The uniqueness follows from the fact that the above construction of ϕ is forced. If ϕ is any diffeomorphism with the required properties, then at $\rho = 0$, $\phi_*(\partial_\lambda)$ is a g-transversal for $\widetilde{g}$, so must be V. Then for $z \in \mathcal{G}$, the curve $I_z \ni \lambda \mapsto \phi(z, \lambda)$ must be the unique geodesic satisfying the initial conditions (2.4). These requirements uniquely determine ϕ on $\mathcal{U}$. $\square$

Chapter Three

Formal Theory

The first goal of this chapter is to prove Theorem 2.9 for $n > 2$. We begin with the following lemma.

Lemma 3.1. *Let $(\widetilde{\mathcal{G}}, \widetilde{g})$ be a pre-ambient space for $(M, [g])$, where $\widetilde{\mathcal{G}}$ has the property that for each $z \in \mathcal{G}$, the set of all $\rho \in \mathbb{R}$ such that $(z, \rho) \in \widetilde{\mathcal{G}}$ is an open interval I_z containing* 0. *Let g be a metric in the conformal class, with associated identification $\mathbb{R}_+ \times M \times \mathbb{R} \simeq \mathcal{G} \times \mathbb{R}$. Then $(\widetilde{\mathcal{G}}, \widetilde{g})$ is in normal form relative to g if and only if one has on $\widetilde{\mathcal{G}}$*

$$\widetilde{g}_{0\infty} = t, \qquad \widetilde{g}_{i\infty} = 0, \qquad \widetilde{g}_{\infty\infty} = 0. \tag{3.1}$$

Proof. Since a pre-ambient metric satisfies $\iota^* \widetilde{g} = \mathbf{g}_0$, if $\widetilde{g}$ satisfies (3.1), then it has the form (2.2) at $\rho = 0$. Thus we must show that for $\widetilde{g}$ satisfying (2.2) at $\rho = 0$, the condition that the lines $\rho \ni I_{(t,x)} \to (t, x, \rho)$ are geodesics for $\widetilde{g}$ is equivalent to (3.1). Now the ρ-lines are geodesics if and only if $\widetilde{\Gamma}_{\infty\infty I} = 0$, where $\widetilde{\Gamma}_{IJK} = \widetilde{g}_{KL} \widetilde{\Gamma}^L_{IJ}$ and $\widetilde{\Gamma}^L_{IJ}$ are the usual Christoffel symbols for $\widetilde{g}$. Taking $I = \infty$ gives $\partial_\rho \widetilde{g}_{\infty\infty} = 0$, which combined with $\widetilde{g}_{\infty\infty}|_{\rho=0} = 0$ from (2.2) yields $\widetilde{g}_{\infty\infty} = 0$. Now taking $I = i$ and $I = 0$ and using (2.2) gives $\widetilde{g}_{\infty i} = 0$ and $\widetilde{g}_{\infty 0} = t$. □

The case $n = 2$ is exceptional for Theorem 2.9. We give the proof for $n > 2$ now; a sharpened version for $n = 2$ will be given in Theorem 3.7.

Proof of Theorem 2.9 for $n > 2$. Given M and a metric g on M, we must construct a smooth metric $\widetilde{g}$ on a suitable neighborhood $\widetilde{\mathcal{G}}$ of $\mathbb{R}_+ \times M \times \{0\}$ with the following properties:

(1) $\delta_s^* \widetilde{g} = s^2 \widetilde{g}, \quad s > 0$;

(2) $\widetilde{g} = t^2 g(x) + 2t\, dt\, d\rho$ when $\rho = 0$;

(3) For each (t, x), the curve $\rho \to (t, x, \rho)$ is a geodesic for $\widetilde{g}$;

(4) If n is odd, then $\mathrm{Ric}(\widetilde{g}) = 0$ to infinite order at $\rho = 0$.

(4′) If n is even, then $\mathrm{Ric}(\widetilde{g}) = O^+_{IJ}(\rho^{n/2-1})$.

Also we must show that if n is odd, then such a metric is uniquely determined to infinite order at $\rho = 0$, while if n is even, then it is determined modulo $O^+_{IJ}(\rho^{n/2})$.

Lemma 3.1 enables us to replace (3) above with (3.1). Thus the components $\widetilde{g}_{I\infty}$ are determined. We consider the remaining components of $\widetilde{g}$ as unknowns, subject to the homogeneity conditions determined by (1) above. Then (2) can be interpreted as initial conditions and the equation $\mathrm{Ric}(\widetilde{g}) = 0$ as a system of partial differential equations to be solved formally.

Set $\widetilde{g}_{00} = a$, $\widetilde{g}_{0i} = tb_i$ and $\widetilde{g}_{ij} = t^2 g_{ij}$, where all of a, b_i, g_{ij} are functions of (x, ρ). Condition (2) gives at $\rho = 0$: $a = 0$, $b_i = 0$ and g_{ij} is the given metric. In order to determine the first derivatives of a, b_i, g_{ij} at $\rho = 0$, we calculate at $\rho = 0$ the components $\widetilde{R}_{IJ} = \mathrm{Ric}_{IJ}(\widetilde{g})$ for $I, J \neq \infty$. This is straightforward but tedious. We have

$$\widetilde{g}^{IJ} = \begin{pmatrix} 0 & 0 & t^{-1} \\ 0 & t^{-2}g^{ij} & -t^{-2}b^i \\ t^{-1} & -t^{-2}b^j & t^{-2}(b_k b^k - a) \end{pmatrix} \tag{3.2}$$

and in particular at $\rho = 0$ we have

$$\widetilde{g}^{IJ} = \begin{pmatrix} 0 & 0 & t^{-1} \\ 0 & t^{-2}g^{ij} & 0 \\ t^{-1} & 0 & 0 \end{pmatrix}.$$

The Christoffel symbols are given by

$$\begin{aligned} 2\widetilde{\Gamma}_{IJ0} &= \begin{pmatrix} 0 & \partial_j a & \partial_\rho a \\ \partial_i a & t(\partial_j b_i + \partial_i b_j - 2g_{ij}) & t\partial_\rho b_i \\ \partial_\rho a & t\partial_\rho b_j & 0 \end{pmatrix} \\ 2\widetilde{\Gamma}_{IJk} &= \begin{pmatrix} 2b_k - \partial_k a & t(\partial_j b_k - \partial_k b_j + 2g_{jk}) & t\partial_\rho b_k \\ t(\partial_i b_k - \partial_k b_i + 2g_{ik}) & 2t^2\Gamma_{ijk} & t^2\partial_\rho g_{ik} \\ t\partial_\rho b_k & t^2\partial_\rho g_{jk} & 0 \end{pmatrix} \\ 2\widetilde{\Gamma}_{IJ\infty} &= \begin{pmatrix} 2 - \partial_\rho a & -t\partial_\rho b_j & 0 \\ -t\partial_\rho b_i & -t^2\partial_\rho g_{ij} & 0 \\ 0 & 0 & 0 \end{pmatrix}, \end{aligned} \tag{3.3}$$

where in the second equation the Γ_{ijk} refers to the Christoffel symbol of the metric g_{ij} with ρ fixed. The Ricci curvature is given by

$$\begin{aligned} \widetilde{R}_{IJ} = \tfrac{1}{2}\widetilde{g}^{KL}\left(\partial^2_{IL}\widetilde{g}_{JK} + \partial^2_{JK}\widetilde{g}_{IL} - \partial^2_{KL}\widetilde{g}_{IJ} - \partial^2_{IJ}\widetilde{g}_{KL}\right) \\ + \widetilde{g}^{KL}\widetilde{g}^{PQ}\left(\widetilde{\Gamma}_{ILP}\widetilde{\Gamma}_{JKQ} - \widetilde{\Gamma}_{IJP}\widetilde{\Gamma}_{KLQ}\right). \end{aligned} \tag{3.4}$$

Computing this using the above gives at $\rho = 0$

$$\widetilde{R}_{00} = \frac{n}{2t^2}(2 - \partial_\rho a),$$

$$\widetilde{R}_{i0} = \frac{1}{2t}\left(\partial^2_{i\rho}a - n\partial_\rho b_i\right), \tag{3.5}$$

$$\widetilde{R}_{ij} = \tfrac{1}{2}\left[(\partial_\rho a - n)\partial_\rho g_{ij} - g^{kl}\partial_\rho g_{kl}g_{ij} + \nabla_i\partial_\rho b_j + \nabla_j\partial_\rho b_i - \partial_\rho b_i\partial_\rho b_j\right] + R_{ij},$$

where in the last equation R_{ij} refers to the Ricci curvature of the initial metric and ∇_i denotes the covariant derivative with respect to its Levi-Civita connection. Setting these to 0 successively shows that the vanishing of these components of $\widetilde{R}_{IJ}$ at $\rho = 0$ is equivalent to the conditions

$$a = 2\rho + O(\rho^2), \qquad b_i = O(\rho^2),$$
$$g_{ij}(x,\rho) = g_{ij}(x) + 2P_{ij}(x)\rho + O(\rho^2), \tag{3.6}$$

where

$$P_{ij} = (n-2)^{-1}\left(R_{ij} - \frac{R}{2(n-1)}g_{ij}\right). \tag{3.7}$$

Next we carry out an inductive perturbation calculation for higher orders. Suppose for some $m \geq 2$ that $\widetilde{g}^{(m-1)}_{IJ}$ is a metric satisfying (3.1), (3.6). Set $\widetilde{g}^{(m)}_{IJ} = \widetilde{g}^{(m-1)}_{IJ} + \Phi_{IJ}$, where

$$\Phi_{IJ} = \rho^m \begin{pmatrix} \phi_{00} & t\phi_{0j} & 0 \\ t\phi_{i0} & t^2\phi_{ij} & 0 \\ 0 & 0 & 0 \end{pmatrix} \tag{3.8}$$

and the ϕ_{IJ} are functions of (x,ρ). From (3.4) it follows that

$$\begin{aligned} \widetilde{R}^{(m)}_{IJ} &= \widetilde{R}^{(m-1)}_{IJ} + \tfrac{1}{2}\widetilde{g}^{KL}\left(\partial^2_{IL}\Phi_{JK} + \partial^2_{JK}\Phi_{IL} - \partial^2_{KL}\Phi_{IJ} - \partial^2_{IJ}\Phi_{KL}\right) \\ &\quad + \widetilde{g}^{KL}\widetilde{g}^{PQ}\left(\widetilde{\Gamma}_{ILP}\Gamma^{\Phi}_{JKQ} + \Gamma^{\Phi}_{ILP}\widetilde{\Gamma}_{JKQ} - \widetilde{\Gamma}_{IJP}\Gamma^{\Phi}_{KLQ} - \Gamma^{\Phi}_{IJP}\widetilde{\Gamma}_{KLQ}\right) \\ &\quad + O(\rho^m), \end{aligned} \tag{3.9}$$

where $\widetilde{g}^{AB}$ and $\widetilde{\Gamma}_{ABC}$ refer to the metric $\widetilde{g}^{(m)}_{IJ}$, and $2\Gamma^{\Phi}_{IJK} = \partial_J\Phi_{IK} + \partial_I\Phi_{JK} - \partial_K\Phi_{IJ}$. These are given modulo $O(\rho^m)$ by

$$2\Gamma^{\Phi}_{IJ0} = \begin{pmatrix} 0 & 0 & \partial_\rho\Phi_{00} \\ 0 & 0 & \partial_\rho\Phi_{i0} \\ \partial_\rho\Phi_{00} & \partial_\rho\Phi_{0j} & 0 \end{pmatrix}$$

$$2\Gamma^{\Phi}_{IJk} = \begin{pmatrix} 0 & 0 & \partial_\rho\Phi_{0k} \\ 0 & 0 & \partial_\rho\Phi_{ik} \\ \partial_\rho\Phi_{0k} & \partial_\rho\Phi_{jk} & 0 \end{pmatrix} \tag{3.10}$$

$$2\Gamma^{\Phi}_{IJ\infty} = \begin{pmatrix} -\partial_\rho\Phi_{00} & -\partial_\rho\Phi_{0j} & 0 \\ -\partial_\rho\Phi_{i0} & -\partial_\rho\Phi_{ij} & 0 \\ 0 & 0 & 0 \end{pmatrix}.$$

On the right-hand side of (3.9), one can take for $\widetilde{g}^{AB}$ and $\widetilde{\Gamma}_{ABC}$ the quantities obtained by substituting (3.6) into (3.2), (3.3). Calculating, one finds

$$
\begin{aligned}
t^2\widetilde{R}^{(m)}_{00} &= t^2\widetilde{R}^{(m-1)}_{00} + m(m-1-\tfrac{n}{2})\rho^{m-1}\phi_{00} + O(\rho^m)\\
t\widetilde{R}^{(m)}_{0i} &= t\widetilde{R}^{(m-1)}_{0i} + m(m-1-\tfrac{n}{2})\rho^{m-1}\phi_{0i} + \tfrac{m}{2}\rho^{m-1}\partial_i\phi_{00} + O(\rho^m)\\
\widetilde{R}^{(m)}_{ij} &= \widetilde{R}^{(m-1)}_{ij} + m\rho^{m-1}\Big[(m-\tfrac{n}{2})\phi_{ij} - \tfrac{1}{2}g^{kl}\phi_{kl}g_{ij}\\
&\qquad + \tfrac{1}{2}(\nabla_j\phi_{0i} + \nabla_i\phi_{0j}) + P_{ij}\phi_{00}\Big] + O(\rho^m)\\
t\widetilde{R}^{(m)}_{0\infty} &= t\widetilde{R}^{(m-1)}_{0\infty} + \tfrac{1}{2}m(m-1)\rho^{m-2}\phi_{00} + O(\rho^{m-1})\\
\widetilde{R}^{(m)}_{i\infty} &= \widetilde{R}^{(m-1)}_{i\infty} + \tfrac{1}{2}m(m-1)\rho^{m-2}\phi_{0i} + O(\rho^{m-1})\\
\widetilde{R}^{(m)}_{\infty\infty} &= \widetilde{R}^{(m-1)}_{\infty\infty} - \tfrac{1}{2}m(m-1)\rho^{m-2}g^{kl}\phi_{kl} + O(\rho^{m-1}).
\end{aligned}
\tag{3.11}
$$

We first consider only the components $\widetilde{R}_{IJ}$ with $I,\ J \neq \infty$. Suppose inductively that $\widetilde{g}^{(m-1)}$ has been determined so that $\mathrm{Ric}_{IJ}(\widetilde{g}^{(m-1)}) = O(\rho^{m-1})$ for $I, J \neq \infty$, and that $\widetilde{g}^{(m-1)}$ is uniquely determined modulo $O(\rho^m)$ by this condition and (3.1). Define $\widetilde{g}^{(m)}$ as above. If n is odd or if n is even and $m \leq n/2$, then the coefficient $m-1-n/2$ appearing in the first two formulae of (3.11) does not vanish, so one can uniquely choose ϕ_{00} and ϕ_{0i} at $\rho = 0$ to make $\widetilde{R}^{(m)}_{00}$ and $\widetilde{R}^{(m)}_{0i}$ be $O(\rho^m)$. The map $\phi_{ij} \to (m-n/2)\phi_{ij} - \frac{1}{2}g^{kl}\phi_{kl}g_{ij}$ is bijective on symmetric 2-tensors unless $m = n/2$ or $m = n$, so except for these m one can similarly make $\widetilde{R}^{(m)}_{ij} = O(\rho^m)$. Thus the induction proceeds up to $m < n/2$ for n even and up to $m < n$ for n odd. Consider the next value of m in each case. For n even and $m = n/2$, one can uniquely determine ϕ_{00} and ϕ_{0i} at $\rho = 0$ to make $\widetilde{R}_{00}$, $\widetilde{R}_{0i} = O(\rho^{n/2})$ just as before. In addition, one can choose $g^{ij}\phi_{ij}$ to guarantee that $g^{ij}\widetilde{R}_{ij} = O(\rho^{n/2})$. Thus for n even, we deduce that $\widetilde{g}_{IJ} \mod O^+_{IJ}(\rho^{n/2})$ is uniquely determined by the condition $\widetilde{R}_{IJ} = O^+_{IJ}(\rho^{n/2-1})$ for $I,\ J \neq \infty$. For n odd and $m = n$, one can again uniquely determine ϕ_{00} and ϕ_{0i} at $\rho = 0$ to make $\widetilde{R}_{00}$, $\widetilde{R}_{0i} = O(\rho^n)$, and now can uniquely determine the trace-free part of ϕ_{ij} to guarantee that $\widetilde{R}_{ij} = \lambda g_{ij}\rho^{n-1} \mod O(\rho^n)$ for some function λ.

In order to analyze the remaining components $\widetilde{R}_{I\infty}$ and to complete the analysis above in the case $m = n$ when n is odd, we consider the contracted Bianchi identity. The Ricci curvature of $\widetilde{g}$ satisfies the Bianchi identity $\widetilde{g}^{JK}\widetilde{\nabla}_I\widetilde{R}_{JK} = 2\widetilde{g}^{JK}\widetilde{\nabla}_J\widetilde{R}_{IK}$. Writing this in terms of coordinate derivatives gives

$$2\widetilde{g}^{JK}\partial_J\widetilde{R}_{IK} - \widetilde{g}^{JK}\partial_I\widetilde{R}_{JK} - 2\widetilde{g}^{JK}\widetilde{g}^{PQ}\widetilde{\Gamma}_{JKP}\widetilde{R}_{QI} = 0. \tag{3.12}$$

Suppose for some $m \geq 2$ that $\widetilde{R}_{IJ} = O(\rho^{m-1})$ for $I,\ J \neq \infty$ and $\widetilde{R}_{I\infty} = O(\rho^{m-2})$. Write out (3.12) for $I = 0, i, \infty$, using (3.2), (3.3), (3.6) and the homogeneity of the components $\widetilde{R}_{IJ}$. Calculating mod $O(\rho^{m-1})$, one obtains

$$\begin{aligned}(n-2-2\rho\partial_\rho)\widetilde{R}_{0\infty}+t\partial_\rho\widetilde{R}_{00}&=O(\rho^{m-1})\\(n-2-2\rho\partial_\rho)\widetilde{R}_{i\infty}-t\partial_i\widetilde{R}_{0\infty}+t\partial_\rho\widetilde{R}_{i0}&=O(\rho^{m-1})\\(n-2-\rho\partial_\rho)\widetilde{R}_{\infty\infty}+g^{jk}\nabla_j\widetilde{R}_{k\infty}+tP_k{}^k\widetilde{R}_{0\infty}-\tfrac{1}{2}g^{jk}\partial_\rho\widetilde{R}_{jk}&=O(\rho^{m-1}).\end{aligned}\tag{3.13}$$

Suppose first that n is even. Let $\widetilde{g}_{IJ}$ be the metric determined above mod $O^+_{IJ}(\rho^{n/2})$ by the requirement $\widetilde{R}_{IJ}=O^+_{IJ}(\rho^{n/2-1})$ for $I,\ J\neq\infty$. We show by induction on m that $\widetilde{R}_{I\infty}=O(\rho^{m-1})$ for $1\le m\le n/2$. The statement is clearly true for $m=1$. Suppose it holds for $m-1$ and write $\widetilde{R}_{I\infty}=\gamma_I\rho^{m-2}$. The hypotheses for (3.13) are satisfied. The first equation of (3.13) gives $(n+2-2m)\gamma_0=O(\rho)$, so $\widetilde{R}_{0\infty}=O(\rho^{m-1})$. The second equation of (3.13) then gives $(n+2-2m)\gamma_i=O(\rho)$, so $\widetilde{R}_{i\infty}=O(\rho^{m-1})$. The last equation then gives $(n-m)\gamma_\infty=O(\rho)$, so $\widetilde{R}_{\infty\infty}=O(\rho^{m-1})$, completing the induction.

Hence, if n is even, we have uniquely determined $\widetilde{g}_{IJ}$ mod $O^+_{IJ}(\rho^{n/2})$ so that $\mathrm{Ric}(\widetilde{g})=O^+_{IJ}(\rho^{n/2-1})$. A finite order Taylor polynomial for $\widetilde{g}$ will be nondegenerate on a neighborhood $\widetilde{\mathcal{G}}$ of $\mathbb{R}_+\times M\times\{0\}$ satisfying the required properties. This concludes the proof of Theorem 2.9 for n even.

If n is odd, let $\widetilde{g}_{IJ}$ be the metric determined above by the requirements $\widetilde{R}_{00},\ \widetilde{R}_{0i}=O(\rho^n)$ and $\widetilde{R}_{ij}=\lambda g_{ij}\rho^{n-1}+O(\rho^n)$. Then $\widetilde{g}_{IJ}$ is uniquely determined up to $O(\rho^{n+1})$ except that $\widetilde{g}_{ij}$ has an additional indeterminacy of the form $cg_{ij}\rho^n$ for some function c. As for n even, consider the induction based on (3.13). The constants $n+2-2m$ never vanish for n odd and m integral, and one may proceed with the induction for all three equations to conclude that $\widetilde{R}_{I\infty}=O(\rho^{n-2})$. The hypotheses of (3.13) now hold with $m=n$. The first two equations give $\widetilde{R}_{0\infty},\ \widetilde{R}_{i\infty}=O(\rho^{n-1})$. In the third equation, the coefficient of $\gamma_\infty\rho^{n-2}$ is now 0, so the third equation reduces to $\lambda\rho^{n-2}=O(\rho^{n-1})$. We conclude that $\lambda=O(\rho)$, i.e., $\widetilde{R}_{ij}=O(\rho^n)$. However, there is still the indeterminacy of $cg_{ij}\rho^n$ in $\widetilde{g}_{ij}$ and we do not yet know that $\widetilde{R}_{\infty\infty}=O(\rho^{n-1})$. These can be dealt with simultaneously by observing directly that one can uniquely choose c at $\rho=0$ to make $\widetilde{R}_{\infty\infty}=O(\rho^{n-1})$. Namely, set $\widetilde{g}'_{IJ}=\widetilde{g}_{IJ}+\Phi_{IJ}$, where Φ_{IJ} is given by (3.8) with $m=n$ and $\phi_{00}=\phi_{0i}=0,\ \phi_{ij}=cg_{ij}$. According to the last formula of (3.11), one has $\widetilde{R}'_{\infty\infty}=\widetilde{R}_{\infty\infty}-\frac{1}{2}n^2(n-1)c\rho^{n-2}+O(\rho^{n-1})$. Therefore, c mod $O(\rho)$ is uniquely determined by the requirement $\widetilde{R}'_{\infty\infty}=O(\rho^{n-1})$. Removing the $'$, we thus have $\widetilde{g}_{IJ}$ uniquely determined mod $O(\rho^{n+1})$ by the conditions $\widetilde{R}_{IJ}=O(\rho^n)$ for $I,\ J\neq\infty$ and $\widetilde{R}_{I\infty}=O(\rho^{n-1})$. We can now proceed inductively to all higher orders with no problems: (3.11) shows that the requirement $\widetilde{R}_{IJ}=0$ to infinite order for $I,\ J\neq\infty$ uniquely determines $\widetilde{g}_{IJ}$ to infinite order, and (3.13) shows that the so-determined $\widetilde{g}_{IJ}$ also satisfies

$\widetilde{R}_{I\infty} = 0$ to infinite order.

Summarizing, the components $\widetilde{g}_{I\infty}$ are given by (3.1). We have determined all derivatives of the components $\widetilde{g}_{IJ}$ for I, $J \neq \infty$ at $\rho = 0$ uniquely to ensure that all components of $\mathrm{Ric}(\widetilde{g})$ vanish to infinite order. By Borel's Theorem, we can find a homogeneous symmetric 2-tensor on a neighborhood of $\rho = 0$ with the prescribed Taylor expansion. We can then choose a dilation-invariant subneighborhood $\widetilde{\mathcal{G}}$ satisfying condition (1) of Definition 2.7 on which this tensor is nondegenerate with signature $(p+1, q+1)$. □

Next we show that the metric $\widetilde{g}_{IJ}$ of Theorem 2.9 takes a special form.

Lemma 3.2. *Let $n \geq 2$. If $\widetilde{g}$ has the form*

$$\widetilde{g}_{IJ} = \begin{pmatrix} 2\rho & 0 & t \\ 0 & t^2 g_{ij} & 0 \\ t & 0 & 0 \end{pmatrix}, \tag{3.14}$$

where $g_{ij} = g_{ij}(x, \rho)$ is a one-parameter family of metrics on M, then the Ricci curvature of $\widetilde{g}$ satisfies $\widetilde{R}_{0I} = 0$.

Proof. For $\widetilde{g}$ of the form (3.14), the Christoffel symbols (3.3) become

$$\widetilde{\Gamma}_{IJ0} = \begin{pmatrix} 0 & 0 & 1 \\ 0 & -tg_{ij} & 0 \\ 1 & 0 & 0 \end{pmatrix}$$

$$\widetilde{\Gamma}_{IJk} = \begin{pmatrix} 0 & tg_{jk} & 0 \\ tg_{ik} & t^2\Gamma_{ijk} & \frac{1}{2}t^2 g'_{ik} \\ 0 & \frac{1}{2}t^2 g'_{jk} & 0 \end{pmatrix} \tag{3.15}$$

$$\widetilde{\Gamma}_{IJ\infty} = \begin{pmatrix} 0 & 0 & 0 \\ 0 & -\frac{1}{2}t^2 g'_{ij} & 0 \\ 0 & 0 & 0 \end{pmatrix}.$$

Here $'$ denotes ∂_ρ. Lemma 3.2 is a straightforward computation from (3.4) using (3.2) and (3.15). Alternate derivations are given in the comments after Proposition 4.7 and in the proof of Proposition 6.1. □

For future reference we record the raised index version of (3.15):

$$\widetilde{\Gamma}^0_{IJ} = \begin{pmatrix} 0 & 0 & 0 \\ 0 & -\frac{1}{2}tg'_{ij} & 0 \\ 0 & 0 & 0 \end{pmatrix}$$

$$\widetilde{\Gamma}^k_{IJ} = \begin{pmatrix} 0 & t^{-1}\delta_j{}^k & 0 \\ t^{-1}\delta_i{}^k & \Gamma^k_{ij} & \frac{1}{2}g^{kl}g'_{il} \\ 0 & \frac{1}{2}g^{kl}g'_{jl} & 0 \end{pmatrix} \tag{3.16}$$

$$\widetilde{\Gamma}^\infty_{IJ} = \begin{pmatrix} 0 & 0 & t^{-1} \\ 0 & -g_{ij} + \rho g'_{ij} & 0 \\ t^{-1} & 0 & 0 \end{pmatrix}.$$

Proposition 3.3. *Let $n > 2$. The metric $\widetilde{g}_{IJ}$ in Theorem 2.9 satisfies $\widetilde{g}_{00} = 2\rho$ and $\widetilde{g}_{0i} = 0$, to infinite order for n odd, and modulo $O(\rho^{n/2+1})$ for n even.*

Proof. Consider the inductive construction of the metric $\widetilde{g}$ in the proof of Theorem 2.9. At each step of the induction, the perturbation terms ϕ_{00} and ϕ_{0i} at $\rho = 0$ are determined from the first two equations of (3.11). If $\widetilde{g}^{(m-1)}$ is of the form (3.14), then by Lemma 3.2 we have $\widetilde{R}^{(m-1)}_{00}$, $\widetilde{R}^{(m-1)}_{0i} = 0$, so we obtain ϕ_{00}, $\phi_{0i} = O(\rho)$. It therefore follows by induction that $\widetilde{g}_{00} = 2\rho$ and $\widetilde{g}_{0i} = 0$ to all orders for which these components are determined. $\square$

The next result shows that the special form given by Proposition 3.3 can be reinterpreted in terms of the conditions of Proposition 2.4.

Proposition 3.4. *Suppose $n \geq 2$. Let the pre-ambient space $(\widetilde{\mathcal{G}}, \widetilde{g})$ be in normal form relative to a representative metric g. The following conditions are equivalent:*

(1) $\widetilde{g}_{00} = 2\rho$ and $\widetilde{g}_{0i} = 0$.

(2) For each $p \in \widetilde{\mathcal{G}}$, the dilation orbit $s \to \delta_s p$ is a geodesic for $\widetilde{g}$.

(3) $2T \lrcorner \widetilde{g} = d(\|T\|^2)$.

(4) The infinitesimal dilation field T satisfies $\widetilde{\nabla} T = Id$.

Proof. By Lemma 3.1, we have (3.1). The computations leading to (3.3) are valid for all $n \geq 2$, so the Christoffel symbols are given by (3.3). Consider a dilation orbit $s \to \delta_s p$ for $p \in \widetilde{\mathcal{G}}$, given in components by $s \to (st, x, \rho)$. Its tangent vector is a constant multiple of ∂_t. These orbits are therefore geodesics for $\widetilde{g}$ if and only if $\widetilde{\Gamma}_{00I} = 0$. From (3.3) and the initial normalization (2.2), it is easily seen that the condition $\widetilde{\Gamma}_{00I} = 0$ is equivalent to $\widetilde{g}_{00} = 2\rho$, $\widetilde{g}_{0i} = 0$. Therefore (1) is equivalent to (2). If $\widetilde{g}$ is any pre-ambient metric, then (3) states that $2t\widetilde{g}_{I0} = \partial_I (t^2 \widetilde{g}_{00})$, while the condition $\widetilde{\Gamma}_{00I} = 0$ can be written $2\partial_0 \widetilde{g}_{0I} = \partial_I \widetilde{g}_{00}$. Both of these are automatic for $I = 0$ and are easily seen to be equivalent for $I = i, \infty$ upon using the homogeneity of the components $\widetilde{g}_{0I}$. Thus (2) and (3) are equivalent. As for (4), if $\widetilde{\nabla} T = Id$, then $\widetilde{\nabla}_T T = T$, which is a restatement of the geodesic condition. On the other hand, if $\widetilde{g}$ has the form (3.14), then (3.16) gives $\widetilde{\nabla}_I T^J = \delta^J{}_I$. $\square$

Proof of Proposition 2.4. Choose a representative metric g and invoke Proposition 2.8. If $\mathcal{U}$ and ϕ are as in Proposition 2.8, then the pre-ambient space $(\mathcal{U}, \phi^* \widetilde{g})$ is in normal form relative to g. By Proposition 3.4, the conditions (1)-(3) of Proposition 2.4 are equivalent for the metric $\phi^* \widetilde{g}$ on $\mathcal{U}$. But all of these conditions are invariant under ϕ, so they are also equivalent for $\widetilde{g}$ on $\phi(\mathcal{U})$. □

Proof of Proposition 2.6. As pointed out after the statement of Proposition 2.6, we only need to prove the existence of a straight ambient space for $(M, [g])$. Choose a representative metric g. According to Proposition 3.3, we may as well take the ambient metric given by Theorem 2.9 to be of the form (3.14). Then Proposition 3.4 shows that upon choosing $\widetilde{\mathcal{G}}$ suitably, the ambient space $(\widetilde{\mathcal{G}}, \widetilde{g})$ is straight. □

Let us return to consider the formal determination of $\widetilde{g}_{IJ}$. According to Proposition 3.3 and Lemma 3.2, $\widetilde{g}$ may be taken of the form (3.14), and for any such $\widetilde{g}$ the equations $\widetilde{R}_{0I} = 0$ already hold. Therefore $\widetilde{g}_{ij} = t^2 g_{ij}(x,\rho)$ can be regarded as the only "unknown" component. One can calculate the remaining components of $\widetilde{R}_{IJ}$ to obtain explicit equations for $g_{ij}(x,\rho)$. Again calculating from (3.4), one finds

$$\begin{aligned} \widetilde{R}_{ij} &= \rho g''_{ij} - \rho g^{kl} g'_{ik} g'_{jl} + \tfrac{1}{2}\rho g^{kl} g'_{kl} g'_{ij} - (\tfrac{n}{2}-1) g'_{ij} - \tfrac{1}{2} g^{kl} g'_{kl} g_{ij} + R_{ij} \\ \widetilde{R}_{i\infty} &= \tfrac{1}{2} g^{kl}(\nabla_k g'_{il} - \nabla_i g'_{kl}) \\ \widetilde{R}_{\infty\infty} &= -\tfrac{1}{2} g^{kl} g''_{kl} + \tfrac{1}{4} g^{kl} g^{pq} g'_{kp} g'_{lq}. \end{aligned} \tag{3.17}$$

Here R_{ij} and ∇ denote the Ricci curvature and Levi-Civita connection of $g_{ij}(x,\rho)$ with ρ fixed. The Taylor expansion of $g_{ij}(x,\rho)$ can be determined by successively differentiating and evaluating at $\rho = 0$ the equations obtained by setting these expressions to 0. For example, simply evaluating the first equation at $\rho = 0$ recovers the fact that $g'_{ij}|_{\rho=0} = 2P_{ij}$, which we obtained in (3.6). According to the proof of Theorem 2.9, for n even the first equation of (3.17) determines the derivatives $\partial_\rho^m g_{ij}$ for $m < n/2$ and also $g^{ij}\partial_\rho^{n/2} g_{ij}$, and then the second and third equations automatically hold mod $O(\rho^{n/2-1})$. In practice, it is easier to calculate the traces $g^{ij}\partial_\rho^m g_{ij}$ for $m \le n/2$ using the last equation rather than the first. For n odd, the first equation determines the derivatives $\partial_\rho^m g_{ij}$ for $m < n$ as well as the trace-free part of $\partial_\rho^n g_{ij}$. The trace part of the first equation at order n is automatically true. The value of $g^{ij}\partial_\rho^n g_{ij}$ is determined by the third equation, and all higher derivatives are then determined by the first equation.

If the initial metric is Einstein, one can identify explicitly the solution $g_{ij}(x,\rho)$. It is straightforward to check that if the initial metric satisfies $R_{ij} = 2\lambda(n-1)g_{ij}$, then $g_{ij}(x,\rho) = (1+\lambda\rho)^2 g_{ij}(x)$ solves (3.17).

In general it is feasible to carry out the first few iterations by hand. One finds at $\rho = 0$

$$\begin{aligned}\tfrac{1}{2}(n-4)g''_{ij} &= -B_{ij} + (n-4)P_i{}^k P_{jk}, \qquad n \neq 4\\ \tfrac{1}{2}(n-4)(n-6)g'''_{ij} &= B_{ij},{}_k{}^k - 2W_{kijl}B^{kl} - 4(n-6)P_{k(i}B_{j)}{}^k - 4P_k{}^k B_{ij}\\ &\quad +4(n-4)P^{kl}C_{(ij)k},{}_l - 2(n-4)C^k{}_i{}^l C_{ljk} + (n-4)C_i{}^{kl}C_{jkl}\\ &\quad +2(n-4)P^k{}_k,{}_l C_{(ij)}{}^l - 2(n-4)W_{kijl}P^k{}_m P^{ml}, \qquad n \neq 4, 6,\end{aligned} \tag{3.18}$$

where W_{ijkl} is the Weyl tensor,

$$C_{ijk} = P_{ij},{}_k - P_{ik},{}_j$$

is the Cotton tensor, and

$$B_{ij} = C_{ijk},{}^k - P^{kl}W_{kijl}$$

is the Bach tensor. The traces are given by

$$\begin{aligned} g^{ij}g''_{ij} &= 2P_{ij}P^{ij}\\ (n-4)g^{ij}g'''_{ij} &= -8P_{ij}B^{ij}, \qquad n \neq 4.\end{aligned} \tag{3.19}$$

The first equation of (3.19) also holds for $n = 4$ and the second for $n = 6$.

The part of the derivatives depending linearly on curvature can be calculated for all orders. Differentiating the last equation of (3.17) shows that for $m \geq 2$, the trace $g^{ij}\partial_\rho^m g_{ij}|_{\rho=0}$ has vanishing linear part. An easy induction using the derivative of Ricci curvature

$$R'_{ij} = \tfrac{1}{2}(g'_{ik},{}_j{}^k + g'_{jk},{}_i{}^k - g'_{ij},{}_k{}^k - g'_k{}^k,{}_{ij}) \tag{3.20}$$

and the Bianchi identity $P_{ik},{}^k = P_k{}^k,{}_i$ shows that at $\rho = 0$

$$(4-n)(6-n)\cdots(2m-n)\partial_\rho^m g_{ij} = 2\left(\Delta^{m-1}P_{ij} - \Delta^{m-2}P_k{}^k,{}_{ij}\right) + \text{lots} \tag{3.21}$$

for $m \geq 2$ (and $m < n/2$ for n even). Here lots denotes quadratic and higher terms involving fewer derivatives of curvature and our sign convention is $\Delta = \nabla^k\nabla_k$.

A further observation can be made concerning the derivatives of g_{ij} at $\rho = 0$: each of them can be expressed in terms only of Ricci curvature and its covariant derivatives. Terms involving Weyl curvature and its derivatives need not appear.

Proposition 3.5. *Each derivative $\partial_\rho^m g_{ij}|_{\rho=0}$ can be expressed as a linear combination of contractions of Ricci curvature and covariant derivatives of Ricci curvature for the initial metric g. This holds for all m for which these expressions are determined: for $m \geq 1$ for n odd and for $1 \leq m < n/2$ for n even, and also for $g^{ij}\partial_\rho^{n/2} g_{ij}|_{\rho=0}$ for n even.*

Proof. The proof is by induction on m. We already know that $g'_{ij}|_{\rho=0} = 2P_{ij}$. Consider the inductive determination of $\partial^m_\rho g_{ij}|_{\rho=0}$ for $m \geq 2$ by taking the equation obtained by setting the first expression of (3.17) equal to 0, applying ∂^{m-1}_ρ, and setting $\rho = 0$. The first, fourth and fifth terms involve $\partial^m_\rho g_{ij}|_{\rho=0}$. The second and third terms give rise to combinations of contractions of previously determined derivatives which are of the desired form by the induction hypothesis. The fifth term also generates contractions of this form in addition to the term involving $\partial^m_\rho g_{ij}$. The result therefore follows if $\partial^{m-1}_\rho R_{ij}|_{\rho=0}$ can be written only in terms of Ricci curvature and its covariant derivatives of the original metric. The first derivative is given by (3.20). In differentiating (3.20) again, the derivative can fall on g', on g^{-1}, or on the connection. Differentiating a Christoffel symbol shows that for a 1-form η_i which depends on ρ, one has $(\eta_{i},{}_j)' = \eta'{}_{i},{}_j - \Gamma'^k_{ij}\eta_k$, where Γ'^k_{ij} is the tensor $\Gamma'^k_{ij} = \frac{1}{2}g^{kl}(g'_{il},{}_j + g'_{jl},{}_i - g'_{ij},{}_l)$. There is an analogous formula for the covariant derivative of a tensor of higher rank. Iterating such formulae together with the Leibnitz formula and the formula for $g^{ij\prime}$, then setting $\rho = 0$ and applying the induction hypothesis, it is clear that $\partial^{m-1}_\rho R_{ij}|_{\rho=0}$ has the desired form.

If n is odd, the trace $g^{ij}\partial^n_\rho g_{ij}|_{\rho=0}$ is determined from the third line of (3.17) rather than from the first. But it is clear that differentiating the third line and using the induction hypothesis gives rise to an expression of the desired form. □

We remark that it is a consequence of Proposition 3.5 that objects constructed solely out of the tensors $\partial^m_\rho g_{ij}|_{\rho=0}$ also can be written in terms of Ricci curvature and its covariant derivatives. Two examples are the "conformally invariant powers of the Laplacian" of [GJMS] and Branson's Q-curvature ([Br]). It is easily seen from the construction in [GJMS] that the coefficients of the conformally invariant natural operators constructed there can be written in terms of the $\partial^m_\rho g_{ij}|_{\rho=0}$; hence by Proposition 3.5 in terms of the Ricci curvature and its derivatives. For Q-curvature, the same GJMS construction can be used to establish the result following Branson's original definition. Alternatively and more directly, this follows from the characterization of Q-curvature given in [FH].

The equations defining an ambient metric in normal form possess a symmetry under reflection in ρ. Let $R : \mathbb{R}_+ \times M \times \mathbb{R} \to \mathbb{R}_+ \times M \times \mathbb{R}$ be $R(t, x, \rho) = (t, x, -\rho)$. If $\widetilde{g}$ is an ambient metric for $(M, [g])$ in normal form relative to a representative g, then $R^*\widetilde{g}$ is also an ambient metric for $(M, [g])$ but is not in normal form because condition (3) of Definition 2.7 does not hold. However, the following proposition is easily verified.

Proposition 3.6. *If $\widetilde{g}$ is an ambient metric for $(M, [g])$ in normal form relative to g, then $-R^*\widetilde{g}$ is an ambient metric for $(M, [-g])$ in normal form* *relative to* $-g$.

Recall that these ambient metrics are unique up to the order specified in Theorem 2.9(B).

Next we prove a sharpened version of Theorem 2.9 for $n = 2$. We denote by tf the trace-free part with respect to g.

Theorem 3.7. *Let M be a smooth manifold of dimension 2. Let g be a smooth metric on M and h a smooth symmetric 2-tensor on M satisfying $g^{ij}h_{ij} = 0$. Then there is an ambient metric $\widetilde{g}$ for $(M, [g])$ in normal form relative to g which satisfies $\operatorname{tf}(\partial_\rho \widetilde{g}_{ij}|_{\rho=0}) = t^2 h_{ij}$. These conditions uniquely determine $\widetilde{g}$ to infinite order at $\rho = 0$. The metric $\widetilde{g}$ is straight to infinite order if and only if $h_{ij},{}^j = \frac{1}{2}R_{,i}$.*

Proof. The proof begins the same way as the proof of Theorem 2.9 for $n > 2$. The components $\widetilde{g}_{I\infty}$ are determined by Lemma 3.1. The computations leading to (3.5) remain valid. The vanishing of the first two lines of (3.5) is again equivalent to $\partial_\rho a|_{\rho=0} = 2$ and $\partial_\rho b_i|_{\rho=0} = 0$. However when $n = 2$, the coefficient $(\partial_\rho a - n)$ vanishes in the third line of (3.5). For $n = 2$, one always has $2R_{ij} = Rg_{ij}$, so vanishing of the third line is therefore equivalent to

$$g^{kl}\partial_\rho g_{kl}|_{\rho=0} = R. \tag{3.22}$$

Thus the trace of $\partial_\rho g_{ij}|_{\rho=0}$ is determined, but the trace-free part remains undetermined by the Einstein condition. We observe that this information is already enough to prove part (B) of Theorem 2.9 when $n = 2$: we have now shown that a solution $\widetilde{g}$ is uniquely determined modulo $O^+_{IJ}(\rho)$. The prescription $\operatorname{tf}(\partial_\rho g_{ij})|_{\rho=0} = h_{ij}$ fixes the ambiguity in $\partial_\rho g_{ij}|_{\rho=0}$. It is convenient to define P_{ij} by

$$2P_{ij} = h_{ij} + \tfrac{1}{2}Rg_{ij}$$

so that $g_{ij}(x, \rho)$ is still given by the second line of (3.6).

We will consider the inductive determination of the Taylor expansion of $\widetilde{g}$ for higher orders as in the proof of Theorem 2.9 for $n > 2$. The next order is the tricky one. For definiteness, we define $\widetilde{g}^{(1)}$ to be given by (3.14) with $g_{ij}(x, \rho)$ given by the second line of (3.6) (say with the $O(\rho^2)$ term set to 0); i.e., we fix the $O(\rho^2)$ indeterminacy in a and b_i in (3.6) to be 0. With notation as in the proof of Theorem 2.9 above, Lemma 3.2 implies that $\widetilde{R}^{(1)}_{0I} = 0$. Also, $\widetilde{R}^{(1)}_{i\infty}$ is given by the middle line of (3.17). In particular,

$$\widetilde{R}^{(1)}_{i\infty}|_{\rho=0} = P_{ij},{}^j - P_j{}^j{}_{,i} = \tfrac{1}{2}\left(h_{ij},{}^j - \tfrac{1}{2}R_{,i}\right). \tag{3.23}$$

Set $\widetilde{g}^{(2)}_{IJ} = \widetilde{g}^{(1)}_{IJ} + \Phi_{IJ}$ with Φ_{IJ} given by (3.8) with $m = 2$. The perturbed Ricci tensor is given by (3.11). The first line says $\widetilde{R}^{(2)}_{00} = O(\rho^2)$. The fourth line shows that $\widetilde{R}^{(2)}_{0\infty} = O(\rho)$ if and only if $\phi_{00} = O(\rho)$, so we require

$\phi_{00}|_{\rho=0} = 0$. The second line then gives $\widetilde{R}^{(2)}_{0i} = O(\rho^2)$. The fifth line shows that $\phi_{0i}|_{\rho=0}$ can be uniquely chosen so that $\widetilde{R}^{(2)}_{i\infty} = O(\rho)$, and (3.23) shows that the so determined $\phi_{0i}|_{\rho=0}$ vanishes if and only if $h_{ij},{}^j = \frac{1}{2}R_{,i}$. The third line shows that the trace-free part of $\phi_{ij}|_{\rho=0}$ can be uniquely chosen so that

$$\widetilde{R}^{(2)}_{ij} = \lambda\rho g_{ij} + O(\rho^2) \tag{3.24}$$

for some function λ on M. The last line shows that the trace of $\phi_{ij}|_{\rho=0}$ can be uniquely determined so that $\widetilde{R}^{(2)}_{\infty\infty} = O(\rho)$. Thus all $\phi_{IJ}|_{\rho=0}$ have been determined, and all components of $\widetilde{R}^{(2)}_{IJ}$ vanish to the desired orders except for (3.24). Consider now the last line of (3.13) applied to $\widetilde{R}^{(2)}$ with $m = 2$. It reduces to $g^{ij}\partial_\rho \widetilde{R}^{(2)}_{ij} = O(\rho)$. Thus $\lambda = 0$ as desired. Therefore we have shown that one can uniquely determine $\Phi_{IJ} \mod O(\rho^3)$ to make $\widetilde{R}^{(1)}_{IJ} = O(\rho^2)$ for $I,\ J \neq \infty$ and $\widetilde{R}^{(1)}_{I\infty} = O(\rho)$. Moreover, the metric $\widetilde{g}^{(2)}_{IJ}$ so determined is of the form (3.14) $\mod O(\rho^3)$ if and only if $h_{ij},{}^j = \frac{1}{2}R_{,i}$.

Consider now the induction for higher m. The argument proceeds as in the proof of Theorem 2.9 above. The relevant coefficients in the first three lines of (3.11) never vanish for $n = 2$ and $m \geq 3$. Thus the conditions $\widetilde{R}^{(m)}_{IJ} = O(\rho^m)$ for $I,\ J \neq \infty$ uniquely determine $\Phi_{IJ} \mod O(\rho^{m+1})$. Then the three lines of (3.13) successively show that $\widetilde{R}^{(m)}_{I\infty} = O(\rho^{m-1})$. Thus the induction continues to all orders.

If $\widetilde{g}$ is straight to infinite order, then $\widetilde{g}_{0i} = 0$ to infinite order, so by the determination of $\phi_{0i}|_{\rho=0}$ when $m = 2$ noted above, we must have $h_{ij},{}^j = \frac{1}{2}R_{,i}$. Conversely, if $h_{ij},{}^j = \frac{1}{2}R_{,i}$, then we have $\phi_{0i}|_{\rho=0}$ when $m = 2$. The argument of Proposition 3.3 then shows that $\widetilde{g}_{00} = 2\rho$ and $\widetilde{g}_{0i} = 0$ to all higher orders. □

Theorem 2.9 for $n = 2$ is a consequence of Theorem 3.7 and its proof. Part (A) follows upon choosing any h_{ij} satisfying $h_i{}^i = 0$. We already noted that part (B) holds at the end of the first paragraph in the proof of Theorem 3.7 above.

We remark that in the straight case, i.e., when $h_{ij},{}^j = \frac{1}{2}R_{,i}$, the solution $\widetilde{g}$ in Theorem 3.7 can be written explicitly; see Chapter 7.

Observe in Theorem 3.7 that the metric $\widetilde{g}$ may be put into normal form relative to another metric $\widehat{g}$ in the conformal class, giving rise to another trace-free tensor $\widehat{h}$. Since the straightness condition is invariant under diffeomorphisms, h satisfies $h_{ij},{}^j = \frac{1}{2}R_{,i}$ if and only if $\widehat{h}$ satisfies $\widehat{h}_{ij},{}^j = \frac{1}{2}\widehat{R}_{,i}$, where in the latter equation the covariant derivative and scalar curvature are that of $\widehat{g}$. Note also that if g has constant scalar curvature, then $h_{ij} = 0$ satisfies $h_i{}^i = 0$ and $h_{ij},{}^j = \frac{1}{2}R_{,i}$. In the case of definite signature, the uniformization theorem implies that every conformal class $(M, [g])$ on any

2-manifold contains a metric of constant scalar curvature. Thus it follows that for any definite signature metric g, there exists a trace-free h satisfying $h_{ij},{}^{j} = \frac{1}{2}R_{,i}$.

When $n \geq 4$ is even, the existence of formal power series solutions for the ambient metric at order $n/2$ is in general obstructed. The obstruction can be identified as a conformally invariant natural tensor generalizing the Bach tensor in dimension 4, which we call the ambient obstruction tensor and denote $\mathcal{O}_{ij}$. We next define the obstruction tensor and establish its basic properties. Suppose that $n \geq 4$ is even and that $\widetilde{g}$ is an ambient metric for $(M, [g])$. By Theorem 2.3, $\widetilde{g}$ is uniquely determined modulo $O^+_{IJ}(\rho^{n/2})$ up to a homogeneous diffeomorphism of $\widetilde{\mathcal{G}}$ which restricts to the identity on $\mathcal{G}$. Set $Q = \|T\|^2 = \widetilde{g}(T, T)$, where as usual T denotes the infinitesimal dilation. Then Q is a defining function for $\mathcal{G} \times \{0\} \subset \widetilde{\mathcal{G}}$ invariantly associated to $\widetilde{g}$, which is homogeneous of degree 2. (To see that Q is a defining function, one can put $\widetilde{g}$ into normal form, whereupon Proposition 3.3 shows that $Q = 2\rho t^2 \mod O(\rho^{n/2+1})$.) We identify $\mathcal{G}$ with $\mathcal{G} \times \{0\}$ via the inclusion ι. Now $\operatorname{Ric}(\widetilde{g}) = O^+_{IJ}(\rho^{n/2-1})$, so $(Q^{1-n/2} \operatorname{Ric} \widetilde{g})|_{T\mathcal{G}}$ is a tensor field on $\mathcal{G}$, homogeneous of degree $2 - n$, which annihilates T. It therefore defines a symmetric 2-tensor-density on M of weight $2 - n$, which is trace-free. If g is a metric in the conformal class, evaluating this tensor-density at the image of g viewed as a section of $\mathcal{G}$ defines a 2-tensor on M which we denote by $(Q^{1-n/2} \operatorname{Ric} \widetilde{g})|_g$. We define the obstruction tensor of g to be

$$\mathcal{O} = c_n (Q^{1-n/2} \operatorname{Ric} \widetilde{g})|_g, \qquad c_n = (-1)^{n/2-1} \frac{2^{n-2}(n/2-1)!^2}{n-2}. \tag{3.25}$$

For $\widetilde{g}$ in normal form relative to g, this reduces to

$$\mathcal{O}_{ij} = 2^{1-n/2} c_n (\rho^{1-n/2} \widetilde{R}_{ij})|_{\rho=0}.$$

Theorem 3.8. *Let $n \geq 4$ be even. The obstruction tensor $\mathcal{O}_{ij}$ of g is independent of the choice of ambient metric $\widetilde{g}$ and has the following properties:*

(1) $\mathcal{O}$ is a natural tensor invariant of the metric g; i.e., in local coordinates the components of $\mathcal{O}$ are given by universal polynomials in the components of g, g^{-1} and the curvature tensor of g and its covariant derivatives, and can be written just in terms of the Ricci curvature and its covariant derivatives. The expression for $\mathcal{O}_{ij}$ takes the form

$$\begin{aligned} \mathcal{O}_{ij} &= \Delta^{n/2-2} \left(P_{ij},{}_k{}^k - P_k{}^k{}_{,ij} \right) + lots \\ &= (3-n)^{-1} \Delta^{n/2-2} W_{kijl},{}^{kl} + lots, \end{aligned} \tag{3.26}$$

where $\Delta = \nabla^i \nabla_i$ and lots denotes quadratic and higher terms in curvature involving fewer derivatives.

(2) One has

$$\mathcal{O}_i{}^i = 0 \qquad\qquad \mathcal{O}_{ij},{}^j = 0.$$

(3) $\mathcal{O}_{ij}$ is conformally invariant of weight $2-n$; i.e., if $0 < \Omega \in C^\infty(M)$ and $\widehat{g}_{ij} = \Omega^2 g_{ij}$, then $\widehat{\mathcal{O}}_{ij} = \Omega^{2-n}\mathcal{O}_{ij}$.

(4) If g_{ij} is conformal to an Einstein metric, then $\mathcal{O}_{ij} = 0$.

Proof. We can assume that $\widetilde{g}$ is in normal form relative to g. Then $\widetilde{g}$ is unique up to addition of Φ_{IJ} of the form (3.8) with $m = n/2$, where ϕ_{00}, ϕ_{0i}, and $g^{ij}\phi_{ij}$ all vanish at $\rho = 0$. The independence of $(Q^{1-n/2}\operatorname{Ric}\widetilde{g})|_g$ on $\widetilde{g}$ is then an immediate consequence of (3.11).

According to Proposition 3.3, we may take $\widetilde{g}$ to satisfy $\widetilde{g}_{00} = 2\rho$, $\widetilde{g}_{0i} = 0$. Then $\mathcal{O}_{ij}$ may be obtained by setting $\widetilde{R}_{ij} = c_n^{-1}(2\rho)^{n/2-1}\mathcal{O}_{ij} \mod O(\rho^{n/2})$ in (3.17), applying $\partial_\rho^{n/2-1}|_{\rho=0}$, and taking the trace-free part. This shows that $\mathcal{O}_{ij}$ is a natural tensor and Proposition 3.5 shows that it can be written just in terms of Ricci and its derivatives. Its linear part may be calculated using (3.20) and (3.21) to be given by the first line of (3.26). The second line follows from the fact that $W_{kijl},{}^{kl} = (3-n)(P_{ij},{}_k{}^k - P_{ik},{}_j{}^k)$.

We have already observed that $\mathcal{O}_{ij}$ is trace-free. Its conformal invariance follows from its definition in terms of a conformally invariant tensor-density. If g_{ij} is Einstein, we have previously noted that there is a solution for $\widetilde{g}$ to all orders, so $\mathcal{O}_{ij} = 0$.

It only remains to establish that $\mathcal{O}_{ij},{}^j = 0$. This follows from the Bianchi identity as follows. Recall that the hypotheses for (3.13) were that $\widetilde{R}_{IJ} = O(\rho^{m-1})$ for $I, J \neq \infty$ and $\widetilde{R}_{I\infty} = O(\rho^{m-2})$. Our metric $\widetilde{g}$ satisfies these with $m = n/2+1$ except that $c_n\widetilde{R}_{ij} = (2\rho)^{n/2-1}\mathcal{O}_{ij} \mod O(\rho^{n/2})$. If one recalculates the middle line of (3.13) allowing the possibility that $\widetilde{R}_{ij} = O(\rho^{m-2})$ but all other components vanish as before, one finds that there are two extra terms: $g^{jk}\nabla_j\widetilde{R}_{ik} - \frac{1}{2}g^{jk}\nabla_i\widetilde{R}_{jk}$. For $m = n/2+1$, the coefficient of $\widetilde{R}_{i\infty}$ vanishes, so we obtain

$$t(\partial_\rho\widetilde{R}_{i0} - \partial_i\widetilde{R}_{0\infty}) + c'\rho^{n/2-1}\mathcal{O}_{ij},{}^j = O(\rho^{n/2})$$

for some nonzero constant c'. We may as well take $\widetilde{g}$ to have $\widetilde{g}_{00} = 2\rho$, $\widetilde{g}_{i0} = 0$, in which case the first two terms vanish, giving the desired conclusion. □

For $n = 4$, 6 one can calculate $\mathcal{O}_{ij}$ by hand by carrying out the computation indicated in the proof of Theorem 3.8. One obtains the tensor which obstructs the validity of (3.18): $\mathcal{O}_{ij} = B_{ij}$ for $n = 4$ and

$$\begin{aligned}\mathcal{O}_{ij} = B_{ij},{}_k{}^k - 2W_{kijl}B^{kl} - 4P_k{}^kB_{ij} + 8P^{kl}C_{(ij)k},{}_l - 4C^k{}_i{}^lC_{ljk} \\ + 2C_i{}^{kl}C_{jkl} + 4P^k{}_{k},{}_l\, C_{(ij)}{}^l - 4W_{kijl}P^k{}_mP^{ml}\end{aligned}$$

for $n = 6$.

When the obstruction tensor is nonzero, there are no formal power series solutions for $\widetilde{g}$ beyond order $n/2$. However, one can still continue the solution to higher orders by introducing log terms. In this case one is obliged to introduce an indeterminacy and the solution is no longer determined solely by the initial metric. We have already seen this indeterminacy phenomenon when $n = 2$ in Theorem 3.7, although for $n = 2$ there is no obstruction and consequently there are no log terms. When n is odd, there are solutions with expansions involving half-integral powers which also have an indeterminacy at order $n/2$.

We broaden our terminology to encompass such metrics. Recall that in Definition 2.1, we required an ambient metric to be smooth. We now define a *generalized ambient metric* to be a metric $\widetilde{g}$ satisfying all the conditions of Definition 2.1, except that the smoothness condition is relaxed to the requirement that $\widetilde{g}$ be $C^\infty(\widetilde{\mathcal{G}} \setminus \{\rho = 0\}) \cap C^1(\widetilde{\mathcal{G}})$, and in all dimensions we require $\operatorname{Ric}(\widetilde{g})$ to vanish to infinite order along $\mathcal{G} \times \{0\}$ (i.e., all derivatives of all components of $\operatorname{Ric}(\widetilde{g})$ extend continuously to $\rho = 0$ and vanish there). Definition 2.7 and Lemma 3.1 concerning metrics in normal form extend to generalized ambient metrics. Proposition 2.4 and the notion of straightness also extend to generalized ambient metrics.

We now discuss the existence and uniqueness of generalized ambient metrics in normal form, beginning with the case n odd.

Theorem 3.9. *Let M be a smooth manifold of odd dimension n. Suppose given a smooth metric g and a smooth symmetric 2-tensor h on M satisfying $g^{ij}h_{ij} = 0$. Then there exists a generalized ambient metric $\widetilde{g}$ which is in normal form relative to g and whose restriction to either $\widetilde{\mathcal{G}} \cap \{\rho > 0\}$ or $\widetilde{\mathcal{G}} \cap \{\rho < 0\}$ has the form*

$$\widetilde{g}_{IJ} = \psi^{(0)}_{IJ} + \psi^{(1)}_{IJ}|\rho|^{n/2}$$

where $\psi^{(0)}_{IJ}$, $\psi^{(1)}_{IJ}$ extend smoothly up to $\rho = 0$ and $\operatorname{tf}\left(\psi^{(1)}_{ij}|_{\rho=0}\right) = t^2 h_{ij}$. *The Taylor expansions of the $\psi^{(0)}_{IJ}$ and $\psi^{(1)}_{IJ}$ are uniquely determined to infinite order by these conditions, and the solution satisfies $g^{ij}\psi^{(1)}_{ij}|_{\rho=0} = 0$. The metric $\widetilde{g}$ is straight to infinite order if and only if $h_{ij},{}^{j} = 0$.*

Proof. We construct $\widetilde{g}$ separately on $\widetilde{\mathcal{G}} \cap \{\rho > 0\}$ and on $\widetilde{\mathcal{G}} \cap \{\rho < 0\}$. First consider $\{\rho > 0\}$.

Return to the inductive construction of $\widetilde{g}$ in Theorem 2.9. If we pause in that construction at $m = (n-1)/2$, we have $\widetilde{g}^{((n-1)/2)}$ determined uniquely mod $O(\rho^{(n+1)/2})$ by (3.1) and the condition $\operatorname{Ric}_{IJ}(\widetilde{g}^{((n-1)/2)}) = O(\rho^{(n-1)/2})$ for $I, J \neq \infty$. Proposition 3.3 shows that we may as well choose $\widetilde{g}^{((n-1)/2)}$ to be of the form (3.14), in which case we have $\widetilde{R}_{00}^{((n-1)/2)} = 0$, $\widetilde{R}_{0i}^{((n-1)/2)} = 0$, and of course $\widetilde{R}_{ij}^{((n-1)/2)}$ is smooth, i.e., it has no half-integral powers in its expansion. Previously we considered only formal power series solutions, so we next modified $\widetilde{g}$ at order $(n+1)/2$. Now we instead modify $\widetilde{g}$ at order $n/2$: set $\widetilde{g}_{IJ}^{(n/2)} = \widetilde{g}_{IJ}^{((n-1)/2)} + \Phi_{IJ}$, where Φ_{IJ} is of the form (3.8) with $m = n/2$ and each ϕ_{IJ} is asymptotic to a formal power series in $\sqrt{\rho}$. It follows from (3.4) that each component of the Ricci curvature of such a metric is also asymptotic to a formal power series in $\sqrt{\rho}$. (For $n = 3$, the series for the components $\widetilde{R}_{I\infty}$ may in principle also contain a $\rho^{-1/2}$ term arising from the $\rho^{3/2}$ term in $\widetilde{g}$; see below.) Now (3.11) still holds except the $\widetilde{R}_{IJ}^{(m-1)}$ which appear on the right-hand side are now replaced by $\widetilde{R}_{IJ}^{(m-1/2)}$ and the error terms are shifted by $\frac{1}{2}$: the error terms in the first three lines are $O(\rho^{m-1/2})$ and those in the last three lines are $O(\rho^{m-3/2})$. Vanishing of the first three lines of (3.11) gives at $\rho = 0$: $\phi_{00} = 0$, $\phi_{0i} = 0$, $g^{ij}\phi_{ij} = 0$, but the trace-free part of ϕ_{ij} may be chosen arbitrarily. We define $\widetilde{g}^{(n/2)}$ by taking $\phi_{ij} = h_{ij}$ and $\widetilde{g}^{(n/2)}$ of the form (3.14), so that $\widetilde{R}_{00}^{(n/2)} = 0$, $\widetilde{R}_{0i}^{(n/2)} = 0$. Also, $\widetilde{R}_{ij}^{(n/2)}$ is asymptotic to a formal power series in $\sqrt{\rho}$, which by construction satisfies $\widetilde{R}_{ij}^{(n/2)} = O(\rho^{(n-1)/2})$. We now modify $g^{(n/2)}$ by addition of a term (3.8) with $m = (n+1)/2$ to obtain $\widetilde{g}^{((n+1)/2)}$. Once again (3.11) holds with the shifted error terms and superscripts on $\widetilde{R}_{IJ}$. None of the relevant constants which appear in (3.11) vanish for this value of m, so we deduce that ϕ_{00}, ϕ_{0i}, and ϕ_{ij} are all detemined at $\rho = 0$ and once again ϕ_{00} and ϕ_{0i} may be taken to be identically 0. Now $\widetilde{R}_{00}^{((n+1)/2)} = 0$, $\widetilde{R}_{0i}^{((n+1)/2)} = 0$, and $\widetilde{R}_{ij}^{((n+1)/2)}$ is asymptotic to a formal power series in $\sqrt{\rho}$ satisfying $\widetilde{R}_{ij}^{((n+1)/2)} = O(\rho^{n/2})$.

Before proceeding with the induction, consider the components $\widetilde{R}_{I\infty}^{((n+1)/2)}$. We have $\widetilde{R}_{0\infty}^{((n+1)/2)} = 0$ by Lemma 3.2. The components $\widetilde{R}_{i\infty}^{((n+1)/2)}$ and $\widetilde{R}_{\infty\infty}^{((n+1)/2)}$ are given by formal power series in $\sqrt{\rho}$. (When $n = 3$, at first glance it appears from the third line of (3.17) that the $\rho^{3/2}$ term in g_{ij} generates a $\rho^{-1/2}$ term in $\widetilde{R}_{\infty\infty}^{((n+1)/2)}$. However this term has coefficient 0 because $g^{ij}h_{ij} = 0$.) Now it is easily checked that (3.13) holds for $m \in \frac{1}{2}\mathbb{Z}$ and for the $\widetilde{g}$ that we are considering with expansions in $\sqrt{\rho}$, still under the same hypotheses: $\widetilde{R}_{IJ} = O(\rho^{m-1})$ for $I, J \neq \infty$ and $\widetilde{R}_{I\infty} = O(\rho^{m-2})$. (When $n = 3$ one modification is required: the error term in the last line is $O(\rho^{m-1}) + O(\rho^{1/2}\widetilde{R}_{0\infty})$, owing to the $\rho^{3/2}$ term in the expansion of g_{ij}.) If we proceed by induction on the order of vanishing of $\widetilde{R}_{i\infty}^{((n+1)/2)}$ and $\widetilde{R}_{\infty\infty}^{((n+1)/2)}$ in (3.13) similarly to the proof of Theorem 2.9, now using $\widetilde{R}_{0I}^{((n+1)/2)} = 0$ and $\widetilde{R}_{ij}^{((n+1)/2)} = O(\rho^{n/2})$, we find $\widetilde{R}_{i\infty}^{((n+1)/2)}$, $\widetilde{R}_{\infty\infty}^{((n+1)/2)} = O(\rho^{n/2-1})$. We make one further observation about $\widetilde{R}^{((n+1)/2)}_{i\infty}$; namely, $(\rho^{1-n/2}\widetilde{R}^{((n+1)/2)}_{i\infty})|_{\rho=0}$ is a constant multiple of $h_{ij},{}^{j}$. To see this, note that $\widetilde{g}^{((n+1)/2)}$ is of the form (3.14) where $g_{ij} = \eta^{(0)}_{ij} + \eta^{(1)}_{ij}\rho^{n/2}$ for $\eta^{(0)}_{ij}$, $\eta^{(1)}_{ij}$ smooth and $\eta^{(1)}_{ij}|_{\rho=0} = h_{ij}$. The middle line of (3.17) together with the fact that $g^{ij}h_{ij} = 0$ then imply that the $\rho^{n/2-1}$ coefficient in the expansion of $\widetilde{R}^{((n+1)/2)}_{i\infty}$ is a multiple of $h_{ij},{}^{j}$.

Return now to the inductive construction of $\widetilde{g}$. We next define $\widetilde{g}^{(n/2+1)}_{IJ} = \widetilde{g}^{((n+1)/2)}_{IJ} + \Phi_{IJ}$ with Φ_{IJ} given by (3.8) with $m = n/2+1$. The first line of (3.11) (with shifted error) tells us that $\widetilde{R}^{(n/2+1)}_{00} = O(\rho^{(n+1)/2})$ independent of the choice of ϕ_{00}. However, the fourth line of (3.11) tells us that we must choose $\phi_{00} = 0$ at $\rho = 0$ in order to make $\widetilde{R}^{(n/2+1)}_{0\infty} = O(\rho^{(n-1)/2})$. Now the second line of (3.11) says that $\widetilde{R}^{(n/2+1)}_{0i} = O(\rho^{(n+1)/2})$ independent of the choice of ϕ_{0i}. However, the fifth line determines $\phi_{0i}|_{\rho=0}$ by the requirement that $\widetilde{R}^{(n/2+1)}_{i\infty} = O(\rho^{(n-1)/2})$. The third line then determines $\phi_{ij}|_{\rho=0}$ by the requirement that $\widetilde{R}_{ij} = O(\rho^{(n+1)/2})$. Taking $m = (n+1)/2$ in (3.13), we already know that the first two lines hold, and the third line tells us that $\widetilde{R}^{(n/2+1)}_{\infty\infty} = O(\rho^{(n-1)/2})$. Thus the $\phi_{IJ}|_{\rho=0}$ have been uniquely determined and we have $\widetilde{R}^{(n/2+1)}_{IJ} = O(\rho^{(n+1)/2})$ for $I,\ J \neq \infty$ and $\widetilde{R}^{(n/2+1)}_{I\infty} = O(\rho^{(n-1)/2})$. In this determination, we found $\phi_{00}|_{\rho=0} = 0$ and $\phi_{0i}|_{\rho=0}$ was determined by the fifth line of (3.11). By the observation noted above that $(\rho^{1-n/2}\widetilde{R}^{((n+1)/2)}_{i\infty})|_{\rho=0}$ is a constant multiple of $h_{ij},{}^{j}$, we deduce that $\phi_{0i}|_{\rho=0}$ is a constant multiple of $h_{ij},{}^{j}$, and in particular $\phi_{0i}|_{\rho=0}$ vanishes if and only if $h_{ij},{}^{j} = 0$.

Now we modify $\widetilde{g}$ to higher orders successively by induction increasing m by $1/2$ each step. The induction statement is that $\widetilde{g}^{(m)}_{IJ}$ is uniquely determined mod $O(\rho^{m+1/2})$ by the requirement that $\widetilde{R}^{(m)}_{IJ} = O(\rho^{m-1/2})$ for $I,\ J \neq \infty$. We have established above the case $m = n/2+1$. Up through $m = n-1/2$, the induction step follows from the first three lines of (3.11) just as in the proof of Theorem 2.9. Again just as in the proof of Theorem 2.9, we then deduce that $\widetilde{R}^{(n-1/2)}_{I\infty} = O(\rho^{n-2})$ using (3.13).

For $m = n$ we encounter the vanishing of the coefficient of the trace in the third line of (3.11). The same reasoning used in the proof of Theorem 2.9 applies here: the fact that we are increasing m by $1/2$ rather than 1 at each step plays no role in that analysis. The induction for higher m then proceeds as usual, including the induction based on (3.13) for the $\widetilde{R}_{I\infty}$ components. Thus it follows that there is a unique series for $\widetilde{g}$ of the desired form for which all components of $\mathrm{Ric}(\widetilde{g})$ vanish to infinite order.

Finally we observe that, just as in the proof of Proposition 3.3, once we get beyond $m = n/2+1$ the special form (3.14) is preserved in the induction.

So if $h_{ij},{}^j = 0$, then the solution has $\widetilde{g}_{00} = 2\rho$, $\widetilde{g}_{0i} = 0$ to infinite order.

We construct $\widetilde{g}$ for $\{\rho < 0\}$ by using the reflection R as in Proposition 3.6. If $\widetilde{g}$ is a solution on $\{\rho > 0\}$ with data $-g$, $-h$, then $-R^*\widetilde{g}$ is a solution on $\{\rho < 0\}$ with data g, h and vice versa. That the solutions match to first order at $\rho = 0$ can be checked using (3.6). In fact, Proposition 3.6 and the uniqueness of the expansion imply that the solution obtained by reflection is $C^{(n-1)/2}$ across $\rho = 0$. □

The analogue of Theorem 3.9 for n even is the following.

Theorem 3.10. *Let M be a smooth manifold of even dimension $n \geq 4$. If g is a smooth metric and h a smooth symmetric 2-tensor on M satisfying $g^{ij}h_{ij} = 0$, then there exists a generalized ambient metric $\widetilde{g}$ in normal form relative to g, which has an expansion of the form*

$$\widetilde{g}_{IJ} \sim \sum_{N=0}^{\infty} \widetilde{g}_{IJ}^{(N)} (\rho^{n/2} \log|\rho|)^N \tag{3.27}$$

where the $\widetilde{g}_{IJ}^{(N)}$ are smooth on $\widetilde{\mathcal{G}}$ and $\operatorname{tf}\left(\partial_\rho^{n/2}\widetilde{g}_{ij}^{(0)}\right) = t^2 h_{ij}$ *at $\rho = 0$. The Taylor expansions of the $\widetilde{g}_{IJ}^{(N)}$ are uniquely determined to infinite order by these conditions. The solution $\widetilde{g}$ is smooth (i.e., $\widetilde{g}^{(N)}$ vanishes to infinite order for $N \geq 1$) if and only if the obstruction tensor $\mathcal{O}_{ij}$ vanishes on M. There is a natural pseudo-Riemannian invariant 1-form D_i such that the solution $\widetilde{g}$ is straight to infinite order if and only if $h_{ij},{}^j = D_i$.*

Remark 3.11. Observe that the corresponding result when $n = 2$ (Theorem 3.7) takes precisely the same form, except that there are no log terms and the obstruction tensor always vanishes. Theorem 3.7 shows that when $n = 2$, one has $D_i = \frac{1}{2}R_{,i}$. For general even n, D_i is given by (3.36) below, and is the same tensor which appears in Theorem 3.1 of [GrH2]. When $n = 4$, one has

$$D_i = 4P^{jk}P_{ij,k} - 3P^{jk}P_{jk,i} + 2P_i{}^j P^k{}_{k,j}.$$

Remark 3.12. We do not know whether for any (M, g) the equation $h_{ij},{}^j = D_i$ admits a solution. We have already observed after the proof of Theorem 3.7 that the existence of a solution is a conformally invariant condition and that there is always a solution if g is definite and $n = 2$. For $n \geq 3$, the equation $h_{ij},{}^j = f_i$ always admits a local solution for any smooth 1-form f_i. This follows from the fact that at each point of the cotangent bundle minus the zero section, the symbol of the operator $\operatorname{div} : \odot_0^2 T^*M \to T^*M$ is surjective, so that there is a right parametrix (see Theorem 19.5.2 of [Hö]). This also implies that on a compact manifold, the range of div has finite codimension. For g definite and M compact, the range of div is the L^2 orthogonal complement of the space of global conformal Killing fields. Using this and the explicit formula for D_i above, one can show that there exists a solution to $h_{ij},{}^j = D_i$ if $n = 4$, g is definite, and M is compact.

We prepare to prove Theorem 3.10. Denote by $\mathcal{A}$ the space of formal asymptotic expansions of scalar functions f on $\mathbb{R}_+ \times M \times \mathbb{R}$ of the form

$$f \sim \sum_{N \geq 0} f^{(N)}(\rho^{n/2} \log |\rho|)^N$$

where each $f^{(N)}$ is smooth and homogeneous of degree 0 in t. It is easily checked that $\mathcal{A}$ is an algebra, that $\mathcal{A}$ is preserved by ∂_{x^i} and $\rho\partial_\rho$, and that $f^{-1} \in \mathcal{A}$ if $f \in \mathcal{A}$ and $f \neq 0$ at $\rho = 0$. The metric $\widetilde{g}$ which we construct actually has a more refined expansion than (3.27). Let $\mathcal{M}$ denote the space of formal asymptotic expansions of metrics $\widetilde{g}$ on $\mathbb{R}_+ \times M \times \mathbb{R}$ of the form

$$\widetilde{g}_{IJ} = \begin{pmatrix} 2\rho + \alpha & tb_j & t \\ tb_i & t^2 g_{ij} & 0 \\ t & 0 & 0 \end{pmatrix} \tag{3.28}$$

with $\alpha \in \rho^{n/2+2}\mathcal{A}$, $b_i \in \rho^{n/2+1}\mathcal{A}$, $g_{ij} \in \mathcal{A}$. We will show that there is a unique expansion $\widetilde{g} \in \mathcal{M}$ satisfying that g_{ij} is the given representative at $\rho = 0$, that $\operatorname{tf}(\partial_\rho^{n/2} g_{ij}^{(0)}) = h_{ij}$ at $\rho = 0$, and that $\operatorname{Ric}(\widetilde{g}) = 0$.

For $\widetilde{g}_{IJ}$ of the form (3.28), the inverse $\widetilde{g}^{IJ}$ and Christoffel symbols $\widetilde{\Gamma}_{IJK}$ are given by (3.2) and (3.3) with $a = 2\rho + \alpha$. Observe that for all IJ we have $t^{2-\#(IJ)}\widetilde{g}^{IJ} \in \mathcal{A}$, where $\#(IJ)$ denotes the number of zeros in the list IJ. Also, for the off-anti-diagonal elements there is an improvement: $t^{2-\#(IJ)}\widetilde{g}^{IJ} \in \rho\mathcal{A}$ unless both IJ are between 1 and n or one is 0 and the other ∞. Similarly, for the Christoffel symbols we have: $\rho\widetilde{\Gamma}_{IJK} \in t^{2-\#(IJK)}\mathcal{A}$ for all components, and $\widetilde{\Gamma}_{IJK} \in t^{2-\#(IJK)}\mathcal{A}$ unless two of IJK are between 1 and n and the third is ∞.

Proposition 3.13. *If $\widetilde{g} \in \mathcal{M}$, then $t^2\widetilde{R}_{00} \in \rho^{n/2+1}\mathcal{A}$, $t\widetilde{R}_{0j}, t\widetilde{R}_{0\infty} \in \rho^{n/2}\mathcal{A}$, $\rho\widetilde{R}_{ij}$, $\rho\widetilde{R}_{i\infty} \in \mathcal{A}$, $\rho^2\widetilde{R}_{\infty\infty} \in \mathcal{A}$.*

Proof. We first derive a formula for the components $\widetilde{R}_{0J}$ for $\widetilde{g}$ of the form (3.28). Set $E_{IJ} = T_{[I,J]}$. The fact that $\mathcal{L}_T\widetilde{g}_{IJ} = 2\widetilde{g}_{IJ}$ implies that $T_{(I,J)} = \widetilde{g}_{IJ}$, so $T_{I,J} = \widetilde{g}_{IJ} + E_{IJ}$. Thus $T_{I,JK} = E_{IJ,K}$. Now $t\widetilde{R}_{0IJK} = T^L\widetilde{R}_{LIJK} = 2T_{I,[JK]} = 2E_{I[J,K]}$, so $t\widetilde{R}_{0J} = t\widetilde{g}^{IK}\widetilde{R}_{0IJK} = \widetilde{g}^{IK}E_{IJ,K}$. Expand the covariant derivative to obtain

$$t\widetilde{R}_{0J} = \widetilde{g}^{IK}\partial_K E_{IJ} - \widetilde{g}^{IK}\widetilde{g}^{PQ}\widetilde{\Gamma}_{IKP}E_{QJ} - \widetilde{g}^{IK}\widetilde{g}^{PQ}\widetilde{\Gamma}_{JKP}E_{IQ}. \tag{3.29}$$

The components of T_I are given by $T_I = t\widetilde{g}_{I0} = (2\rho t + \alpha t, t^2 b_i, t^2)$. The 1-form $2\rho t dt + t^2 d\rho$ is $d(\rho t^2)$, so these terms can be ignored when calculating $E_{IJ} = T_{[I,J]} = \partial_{[J}T_{I]}$. Thus we have

$$2E_{IJ} = \begin{pmatrix} 0 & t(\partial_j\alpha - 2b_j) & t\partial_\rho\alpha \\ t(2b_i - \partial_i\alpha) & t^2(\partial_j b_i - \partial_i b_j) & t^2\partial_\rho b_i \\ -t\partial_\rho\alpha & -t^2\partial_\rho b_j & 0 \end{pmatrix}.$$

We will use that this is of the form

$$E_{IJ} = \begin{pmatrix} 0 & t\rho^{n/2+1}\mathcal{A} & t\rho^{n/2+1}\mathcal{A} \\ t\rho^{n/2+1}\mathcal{A} & t^2\rho^{n/2+1}\mathcal{A} & t^2\rho^{n/2}\mathcal{A} \\ t\rho^{n/2+1}\mathcal{A} & t^2\rho^{n/2}\mathcal{A} & 0 \end{pmatrix}. \tag{3.30}$$

The conclusion of Proposition 3.13 for the components $\widetilde{R}_{0J}$ follows upon expanding the contractions in (3.29) and using (3.30) and our knowledge of $\widetilde{g}^{IJ}$ and $\widetilde{\Gamma}_{IJK}$. We indicate the details for $\widetilde{R}_{00}$; the cases $\widetilde{R}_{0j}$ and $\widetilde{R}_{0\infty}$ are similar. In all cases, one knows ahead of time that the powers of t work out correctly.

Setting $J=0$ in (3.29) gives

$$t^2\widetilde{R}_{00} = t\widetilde{g}^{IK}\partial_K E_{I0} - t\widetilde{g}^{IK}\widetilde{g}^{PQ}\widetilde{\Gamma}_{IKP}E_{Q0} - t\widetilde{g}^{IK}\widetilde{g}^{PQ}\widetilde{\Gamma}_{0KP}E_{IQ}.$$

It follows using $E_{I0} \in t^{1-\#(I)}\rho^{n/2+1}\mathcal{A}$ and $t^{2-\#(IK)}\widetilde{g}^{IK} \in \mathcal{A}$ that when the first term on the right-hand side is expanded, all terms are in $\rho^{n/2+1}\mathcal{A}$ except possibly those with $K=\infty$. Since $E_{00}=0$, only the terms with $K=\infty$ and $I\neq 0$ need be considered. But $t^{2-\#(I)}\widetilde{g}^{I\infty} \in \rho\mathcal{A}$ for $I \neq 0$, so the first term is in $\rho^{n/2+1}\mathcal{A}$.

A term in the expansion of the second term is in $\rho^{n/2+1}\mathcal{A}$ unless $\widetilde{\Gamma}_{IKP} \notin t^{2-\#(IKP)}\mathcal{A}$. This gives that two of IKP must be between 1 and n and the third must be ∞. Each of the pairs IK, PQ must be 0∞ or with both indices between 1 and n. Q cannot be 0. There are no such possibilities, so the second term is in $\rho^{n/2+1}\mathcal{A}$.

For the third term, we have $\widetilde{\Gamma}_{0KP} \in t^{1-\#(KP)}\mathcal{A}$. So for a term not to be in $\rho^{n/2+1}\mathcal{A}$, it must be that E_{IQ} is not in $t^{2-\#(IQ)}\rho^{n/2+1}\mathcal{A}$, which gives that one of IQ is between 1 and n and the other is ∞. Again each of the pairs IK, PQ must be 0∞ or with both indices between 1 and n. So one of KP must be 0 and the other between 1 and n. However, (3.3) shows that $\widetilde{\Gamma}_{0k0}$, $\widetilde{\Gamma}_{00k} \in \rho\mathcal{A}$, so the third term is in $\rho^{n/2+1}\mathcal{A}$.

For the components $\widetilde{R}_{IJ}$ in which neither I nor J is 0, we use (3.4). It is straightforward to check using the observations above about $\widetilde{g}^{IJ}$ and $\widetilde{\Gamma}_{IJK}$ that for each of the possibilities $IJ = ij$, $i\infty$, $\infty\infty$, each term on the right-hand side of (3.4) is in $\mathcal{A}$ when multiplied by the indicated power of ρ. □

In order to carry out the inductive perturbation analysis for Theorem 3.10, we need to extend (3.11) to the case where the perturbations involve log terms. If $0 \leq m \in \mathbb{Z}$, we will say that an expansion is O^m if it can be written in the form $\sum_{N\geq 0} u^{(N)}(\log|\rho|)^N$, where each $u^{(N)}$ is smooth, homogeneous of some degree in t, and $O(\rho^m)$. Set $\mathcal{A}^m = O^m \cap \mathcal{A}$. The same calculations that gave (3.11) give the following.

Proposition 3.14. *Let $\widetilde{g}$ have the form* (3.28) *with* α, $b_i \in \mathcal{A}^2$, $g_{ij}(x,\rho) = g_{ij}(x) + 2P_{ij}\rho + \mathcal{A}^2$. *Set* $\widetilde{g}'_{IJ} = \widetilde{g}_{IJ} + \Phi_{IJ}$, *where*

$$\Phi_{IJ} = \begin{pmatrix} \phi_{00} & t\phi_{0j} & 0 \\ t\phi_{i0} & t^2\phi_{ij} & 0 \\ 0 & 0 & 0 \end{pmatrix} \tag{3.31}$$

with the $\phi_{IJ} \in \mathcal{A}^m$, $m \geq 2$. *Then*

$$\begin{aligned}
t^2\widetilde{R}'_{00} &= t^2\widetilde{R}_{00} + (\rho\partial_\rho^2 - \tfrac{n}{2}\partial_\rho)\phi_{00} + O^m \\
t\widetilde{R}'_{0i} &= t\widetilde{R}_{0i} + (\rho\partial_\rho^2 - \tfrac{n}{2}\partial_\rho)\phi_{0i} + \tfrac{1}{2}\partial^2_{i\rho}\phi_{00} + O^m \\
\widetilde{R}'_{ij} &= \widetilde{R}_{ij} + \left[\rho\partial_\rho^2 + (1-\tfrac{n}{2})\partial_\rho\right]\phi_{ij} - \tfrac{1}{2}g^{kl}\partial_\rho\phi_{kl}g_{ij} \\
&\quad + \tfrac{1}{2}(\nabla_j\partial_\rho\phi_{0i} + \nabla_i\partial_\rho\phi_{0j}) + P_{ij}\partial_\rho\phi_{00} + O^m \\
t\widetilde{R}'_{0\infty} &= t\widetilde{R}_{0\infty} + \tfrac{1}{2}\partial_\rho^2\phi_{00} + O^{m-1} \\
\widetilde{R}'_{i\infty} &= \widetilde{R}_{i\infty} + \tfrac{1}{2}\partial_\rho^2\phi_{0i} + O^{m-1} \\
\widetilde{R}'_{\infty\infty} &= \widetilde{R}_{\infty\infty} - \tfrac{1}{2}g^{kl}\partial_\rho^2\phi_{kl} + O^{m-1}.
\end{aligned} \tag{3.32}$$

We also need the analogue of (3.13) for expansions with logs. The same reasoning as for (3.13) shows that if $\widetilde{g} \in \mathcal{M}$ satisfies for some $m \geq 2$ that $\widetilde{R}_{IJ} = O^{m-1}$ for I, $J \neq \infty$ and $\widetilde{R}_{I\infty} = O^{m-2}$, then (3.13) holds except that the error terms are all O^m rather than $O(\rho^m)$.

Proof of Theorem 3.10. The analysis of Theorem 2.9 and Proposition 3.3 leaves us with a smooth $g_{ij}(x,\rho)$ determined modulo $O(\rho^{n/2})$ so that the metric $\widetilde{g}$ defined by (3.14) satisfies $\widetilde{R}_{0I} = 0$, $\widetilde{R}_{ij}$, $\widetilde{R}_{j\infty} = O(\rho^{n/2-1})$, $\widetilde{R}_{\infty\infty} = O(\rho^{n/2-2})$. (We will save the determination of the trace of the $\rho^{n/2}$ term in the expansion of g_{ij} for the next step of the induction.) Each coefficient in the Taylor expansion of g_{ij} through order $n/2-1$ is a natural tensorial invariant of the initial metric g. Let us write $(\frac{n}{2}-1)!\,\widetilde{R}_{ij} = \rho^{n/2-1}r_{ij}$ and $(\frac{n}{2}-1)!\,\widetilde{R}_{j\infty} = \rho^{n/2-1}r_{j\infty}$, where r_{ij} and $r_{j\infty}$ are smooth and independent of t. The obstruction tensor is given by

$$(\tfrac{n}{2}-1)!\,\mathcal{O}_{ij} = 2^{1-n/2}c_n\,\mathrm{tf}\,(r_{ij})|_{\rho=0}$$

with c_n as in (3.25). The values at $\rho = 0$ of $r_{j\infty}$ and the trace $g^{kl}r_{kl}$ do depend on the choice of g_{ij} at order $n/2$. However, if we fix g_{ij} to be its finite Taylor polynomial of order $n/2-1$, then these values can be expressed as natural tensorial invariants of the initial metric g.

For specificity, fix g_{ij} to be this Taylor polynomial. Define $\widetilde{g}'_{IJ}$ as in Proposition 3.14 with $m = n/2$. It is clear from (3.32) that only the ϕ_{IJ}

mod $\mathcal{A}^{n/2+1}$ can affect the $\widetilde{R}_{IJ}$ at the next order. So we may as well take

$$(\tfrac{n}{2})!\,\phi_{00} = \rho^{n/2}\left(\kappa^{(1)}\log|\rho| + \kappa^{(0)}\right)$$
$$(\tfrac{n}{2})!\,\phi_{0i} = \rho^{n/2}\left(\mu_i^{(1)}\log|\rho| + \mu_i^{(0)}\right)$$
$$(\tfrac{n}{2})!\,\phi_{ij} = \rho^{n/2}\left(\lambda_{ij}^{(1)}\log|\rho| + \lambda_{ij}^{(0)}\right)$$

with coefficients $\kappa^{(N)}$, $\mu_i^{(N)}$, $\lambda_{ij}^{(N)}$, $N = 0, 1$, smooth and independent of t.

The first equation of (3.32) and the requirement $\widetilde{R}'_{00} = O^{n/2}$ give $(\rho\partial_\rho^2 - \frac{n}{2}\partial_\rho)\phi_{00} = O^{n/2}$, which is easily seen to imply that $\phi_{00} = O^{n/2+1}$. Similarly the second equation of (3.32) and the requirement $\widetilde{R}'_{0i} = O^{n/2}$ give $\phi_{0i} = O^{n/2+1}$. The third equation and the requirement $\widetilde{R}'_{ij} = O^{n/2}$ give

$$r_{ij} - \tfrac{1}{2}g^{kl}\lambda_{kl}^{(1)}g_{ij}\log|\rho| - \left(\tfrac{1}{2}g^{kl}\lambda_{kl}^{(0)} + \tfrac{1}{n}g^{kl}\lambda_{kl}^{(1)}\right)g_{ij} + \lambda_{ij}^{(1)} = O^1. \tag{3.33}$$

Clearly we must have $g^{kl}\lambda_{kl}^{(1)} = 0$ at $\rho = 0$. Taking the trace-free part then shows that $\mathrm{tf}(r_{ij}) + \lambda_{ij}^{(1)} = 0$ at $\rho = 0$, so

$$c_n\lambda_{ij}^{(1)}|_{\rho=0} = -2^{n/2-1}(\tfrac{n}{2}-1)!\,\mathcal{O}_{ij}.$$

Now taking the trace in (3.33) gives $g^{kl}r_{kl} = \frac{n}{2}g^{kl}\lambda_{kl}^{(0)}$ at $\rho = 0$. These determinations are necessary and sufficient for $\widetilde{R}'_{ij} = O^{n/2}$. The trace-free part of $\lambda_{ij}^{(0)}$ is undetermined by the Einstein condition, but is fixed by the choice of h_{ij}: $\mathrm{tf}(\lambda_{ij}^{(0)}|_{\rho=0}) = h_{ij}$. Thus

$$\lambda_{ij}^{(0)}|_{\rho=0} = h_{ij} + \frac{2}{n^2}r_k{}^k g_{ij}.$$

Now $\widetilde{g}'_{IJ}$ is determined mod $\mathcal{A}^{n/2+1}$. We fix the $\mathcal{A}^{n/2+1}$ indeterminacy in $\widetilde{g}'_{00}$, $\widetilde{g}'_{0i}$ by taking ϕ_{00}, $\phi_{0i} = 0$, so that $\widetilde{g}'_{IJ}$ has the form (3.14). Then $\widetilde{R}'_{0I} = 0$, $\widetilde{R}'_{ij} = O^{n/2}$, $\widetilde{R}'_{i\infty} = O^{n/2-1}$, $\widetilde{R}'_{\infty\infty} = O^{n/2-2}$. Also we know that $\rho^2\widetilde{R}'_{\infty\infty} \in \mathcal{A}$ by Proposition 3.13. Substituting this information into the last line of (3.13) with $m = n/2$ shows that $\widetilde{R}'_{\infty\infty} = O^{n/2-1}$. For reference in the next step, we will need to know the leading term of $\widetilde{R}'_{i\infty}$. This component is given explicitly by the middle line of (3.17) (where, however, $'$ denotes ∂_ρ). Replacing g_{ij} by $g_{ij} + \phi_{ij}$ in (3.17) and recalling $\mathcal{O}_j{}^j = 0$, $\mathcal{O}_{ij},{}^j = 0$, one finds that

$$(\tfrac{n}{2}-1)!\,\widetilde{R}'_{i\infty} = \rho^{n/2-1}\left(r_{i\infty} + \frac{1}{2}h_{ij},{}^j - \frac{n-1}{n^2}r_j{}^j,{}_i\right) + O^{n/2}. \tag{3.34}$$

This completes the $m = n/2$ step. The metric $\widetilde{g}'_{IJ}$ has the form (3.14) and its Ricci curvature satisfies $\widetilde{R}'_{0I} = 0$, $\widetilde{R}'_{ij} = O^{n/2}$, $\widetilde{R}'_{i\infty}$, $\widetilde{R}'_{\infty\infty} = O^{n/2-1}$.

Now rename what was $\widetilde{g}'_{IJ}$ to be a new $\widetilde{g}_{IJ}$. Proposition 3.13 shows that this new $\widetilde{g}_{IJ}$ has $\rho\widetilde{R}_{ij} \in \mathcal{A}$, so we can write

$$(\tfrac{n}{2})!\,\widetilde{R}_{ij} = \rho^{n/2}\left(r_{ij}^{(1)}\log|\rho| + r_{ij}^{(0)}\right) + O^{n/2+1} \tag{3.35}$$

with coefficients $r_{ij}^{(N)}$, $N = 0, 1$, smooth and independent of t. Now construct a new $\widetilde{g}'_{IJ}$ as in Proposition 3.14 with $m = n/2+1$. This time we take

$$\begin{aligned}
(\tfrac{n}{2}+1)!\;\phi_{00} &= \rho^{n/2+1}\left(\kappa^{(1)}\log|\rho| + \kappa^{(0)}\right)\\
(\tfrac{n}{2}+1)!\;\phi_{0i} &= \rho^{n/2+1}\left(\mu_i^{(1)}\log|\rho| + \mu_i^{(0)}\right)\\
(\tfrac{n}{2}+1)!\;\phi_{ij} &= \rho^{n/2+1}\left(\lambda_{ij}^{(1)}\log|\rho| + \lambda_{ij}^{(0)}\right).
\end{aligned}$$

Referring to (3.32), the requirement $\widetilde{R}'_{00} = O^{n/2+1}$ is equivalent to $\kappa^{(1)}|_{\rho=0} = 0$; no condition is imposed on $\kappa^{(0)}$. However, the requirement $\widetilde{R}'_{0\infty} = O^{n/2}$ is equivalent to $\phi_{00} = O^{n/2+2}$. Thus the $\kappa^{(N)}|_{\rho=0}$ are determined and we may as well take $\phi_{00} = 0$. Similarly, the requirement $\widetilde{R}'_{0i} = O^{n/2+1}$ is equivalent to $\mu_i^{(1)}|_{\rho=0} = 0$, so we take $\mu_i^{(1)} = 0$. The requirement $\widetilde{R}'_{i\infty} = O^{n/2}$ uniquely determines $\mu_i^{(0)}|_{\rho=0}$. Note that according to (3.34), we have $\mu_i^{(0)}|_{\rho=0} = 0$ (and therefore $\phi_{0i} = O^{n/2+2}$) if and only if $h_{ij},{}^j = D_i$, where

$$D_i = 2\left(\frac{n-1}{n^2}r_j{}^j{}_{,i} - r_{i\infty}\right)\Big|_{\rho=0}. \tag{3.36}$$

An easy computation from the third line of (3.32) using (3.35) shows that the requirement $\widetilde{R}'_{ij} = O^{n/2+1}$ uniquely determines the $\lambda_{ij}^{(N)}|_{\rho=0}$. Now the third line of (3.13) with $m = n/2+1$ shows that $\widetilde{R}'_{\infty\infty} = O^{n/2}$. This completes the $m = n/2+1$ step. We have $\widetilde{g}'_{IJ} \in \mathcal{M}$ with $\widetilde{R}'_{00}$, $\widetilde{R}'_{0i}$, $\widetilde{R}'_{ij} = O^{n/2+1}$, $\widetilde{R}'_{I\infty} = O^{n/2}$. Moreover, it always holds that $\widetilde{g}'_{00} = 2\rho + O^{n/2+2}$, and $\widetilde{g}'_{0i} = O^{n/2+2}$ if and only if $h_{ij},{}^j = D_i$. If $\mathcal{O}_{ij} = 0$, then no log terms occur in any of the expansions, and $\widetilde{g}'_{IJ}$ is smooth.

We now prove by induction on m that there is a metric $\widetilde{g} \in \mathcal{M}$, with α, b_i, g_{ij} in (3.28) uniquely determined mod O^m, such that $\widetilde{R}_{IJ} = O^{m-1}$ for $I, J \neq \infty$ and $\widetilde{R}_{I\infty} = O^{m-2}$. We will also show that the metric $\widetilde{g}$ satisfies $\widetilde{g}_{00} = 2\rho + O^m$ and $\widetilde{g}_{0i} = O^m$ if and only if $h_{ij},{}^j = D_i$. Moreover, $\widetilde{g}$ is smooth if and only if $\mathcal{O}_{ij} = 0$. We have established this for $m = n/2+2$.

The argument for the induction step passing from m to $m+1$ differs depending on whether or not $m = n$. First assume $m \neq n$, and of course $m \geq n/2+2$. Proposition 3.13 implies that the Ricci curvature of $\widetilde{g}_{IJ}$ takes the form

$$
\begin{aligned}
(m-1)!\, t^2 \widetilde{R}_{00} &= \rho^{m-1} \sum_{N=0}^{M_1} r_{00}^{(N)} (\log|\rho|)^N + O^m \\
(m-1)!\, t \widetilde{R}_{0i} &= \rho^{m-1} \sum_{N=0}^{M_2} r_{0i}^{(N)} (\log|\rho|)^N + O^m \\
(m-1)! \widetilde{R}_{ij} &= \rho^{m-1} \sum_{N=0}^{M_3} r_{ij}^{(N)} (\log|\rho|)^N + O^m
\end{aligned} \tag{3.37}
$$

where

$$
M_1 = \left\lfloor \frac{2(m-2)}{n} - 1 \right\rfloor, \qquad M_2 = \left\lfloor \frac{2(m-1)}{n} - 1 \right\rfloor, \qquad M_3 = \left\lfloor \frac{2m}{n} \right\rfloor
$$

and the coefficient functions $r_{IJ}^{(N)}$ are smooth and independent of t. Define $\widetilde{g}'_{IJ}$ as in Proposition 3.14. The requirement that $\widetilde{g}' \in \mathcal{M}$ implies that the perturbation terms take the form

$$
\begin{aligned}
m!\, \phi_{00} &= \rho^m \sum_{N=0}^{M_1} \kappa^{(N)} (\log|\rho|)^N \\
m!\, \phi_{0i} &= \rho^m \sum_{N=0}^{M_2} \mu_i^{(N)} (\log|\rho|)^N \\
m!\, \phi_{ij} &= \rho^m \sum_{N=0}^{M_3} \lambda_{ij}^{(N)} (\log|\rho|)^N
\end{aligned} \tag{3.38}
$$

with coefficients smooth and independent of t. (It is straightforward to modify the argument to allow more general perturbations; for example, only to require that $\widetilde{g}'_{00}$, $\widetilde{g}'_{0i} \in \mathcal{A}$. One finds that the only solution is the one constructed here with $\widetilde{g}' \in \mathcal{M}$.) Since $\widetilde{g}' \in \mathcal{M}$, the components $\widetilde{R}'_{00}$, $\widetilde{R}'_{0i}$, $\widetilde{R}'_{ij}$ take the same form as in (3.37) with coefficients $r'^{(N)}_{IJ}$ determined by (3.32). Upon substituting the first line of (3.38) into the first line of (3.32), one finds by considering the coefficients of $(\log|\rho|)^N$ inductively that the requirement $\widetilde{R}'_{00} = O^m$ can be satisfied and uniquely determines the coefficients $\kappa^{(N)}$ at $\rho = 0$. Since $M_1 \leq M_2$, the term $\frac{1}{2}\partial^2_{i\rho}\phi_{00}$ in the second line of (3.32) can be written in the same form as that of $t\widetilde{R}'_{0i}$ in (3.37). Substituting the second line of (3.38) into (3.32), one finds that the requirement $\widetilde{R}'_{0i} = O^m$ can be satisfied and uniquely determines the $\mu_i^{(N)}$ at $\rho = 0$. Similarly, using $m \neq n$, one finds that the requirement $\widetilde{R}'_{ij} = O^m$ can be satisfied and uniquely determines the $\lambda_{ij}^{(N)}$ at $\rho = 0$. Now consider the Bianchi identities (3.13). The form of the components $\widetilde{R}'_{I\infty}$ is given by Proposition 3.13 and the induction hypothesis. Substituting into (3.13) and using the vanishing $\widetilde{R}'_{IJ} = O^m$ for $I, J \neq \infty$, one finds successively that $\widetilde{R}'_{0\infty} = O^{m-1}$, $\widetilde{R}'_{i\infty} = O^{m-1}$, and (again using $m \neq n$) $\widetilde{R}'_{\infty\infty} = O^{m-1}$. It is evident from Lemma 3.2 that if $\widetilde{g}$ has the form (3.14), then $\widetilde{g}'$ does too, and it is also evident that if $\widetilde{g}$ is smooth, then $\widetilde{g}'$ is too. This concludes the induction step in case $m \neq n$.

Finally consider the case $m = n$. The argument for the determination of the components ϕ_{00} and ϕ_{0i} mod O^{n+1} is unchanged. The first two lines of (3.13) then show that $\widetilde{R}'_{0\infty}$, $\widetilde{R}'_{i\infty} = O^{n-1}$. For $m = n$ we have $M_3 = 2$. Substituting into the third line of (3.32), one finds that one can uniquely choose the coefficients $\lambda_{ij}^{(N)}$ for $N = 1, 2$ and the trace-free part $\operatorname{tf}(\lambda_{ij}^{(0)})$ at $\rho = 0$ to make $r_{ij}^{\prime(0)}$, $r_{ij}^{\prime(1)}$, $\operatorname{tf}(r_{ij}^{\prime(2)}) = O(\rho)$. This leaves us with

$$\widetilde{R}'_{ij} = c\rho^{n-1}(\log|\rho|)^2 g_{ij} + O^n$$

for some smooth c, and there remains an indeterminacy of a multiple of $\rho^n g_{ij}$ in ϕ_{ij}. Proposition 3.13 and the induction hypothesis imply that

$$(n-2)!\,\widetilde{R}'_{\infty\infty} = \rho^{n-2}\sum_{N=0}^{2} r_{\infty\infty}^{\prime(N)}(\log|\rho|)^N + O^{n-1}.$$

Substituting this information into the third line of (3.13), one finds that c, $r_{\infty\infty}^{\prime(2)}$, and $r_{\infty\infty}^{\prime(1)}$ all vanish at $\rho = 0$. Thus we have $\widetilde{R}'_{ij} = O^n$ and $(n-2)!\,\widetilde{R}'_{\infty\infty} = \rho^{n-2} r_{\infty\infty}^{\prime(0)} + O^{n-1}$. Now an inspection of the last line of (3.32) shows that one can uniquely fix the $\rho^n g_{ij}$ indeterminacy in ϕ_{ij} to kill this last coefficient in $\widetilde{R}'_{\infty\infty}$, completing the induction step. In principle, this argument allows the possibility that a log term might be created in ϕ_{ij} even if $\widetilde{g}$ is smooth, but the same reasoning as in the proof for n odd in Theorem 2.9 shows that this potential log term does not occur. □

We remark that similar arguments using the form of the perturbation formulae (3.32) for the Ricci curvature show that the metrics constructed in Theorems 3.7, 3.9 and 3.10 are the only formal expansions of metrics for $\rho > 0$ or $\rho < 0$ involving positive powers of $|\rho|$ and $\log|\rho|$ which are homogeneous of degree 2, Ricci-flat to infinite order, and in normal form.

Convergence of formal series determined by Fuchsian problems such as these in the case of real-analytic data has been considered by several authors. In particular, results of [BaoG] can be applied to establish the convergence of the series occurring in Theorems 3.7 and 3.9 (and also in Theorem 3.10 if the obstruction tensor vanishes) if g and h are real-analytic. Convergence results including also the case when log terms occur in Theorem 3.10 are contained in [K].

Other treatments of various aspects of the construction and properties of ambient metrics are contained in [ČG], [GP1], [BrG], [GP2].

Chapter Four

Poincaré Metrics

In this chapter we consider the formal theory for Poincaré metrics associated to a conformal manifold $(M,[g])$. We will see that even Poincaré metrics are in one-to-one correspondence with straight ambient metrics, if both are in normal form. Thus the formal theory for Poincaré metrics is a consequence of the results of Chapter 3. The derivation of a Poincaré metric from an ambient metric was described in [FG], and the inverse construction of an ambient metric as the cone metric over a Poincaré metric was given in §5 of [GrL].

The definition of Poincaré metrics is motivated by the example of the hyperbolic metric $4(1-|x|^2)^{-2}g_e$ on the ball, where g_e denotes the Euclidean metric. Let $(M,[g])$ be a smooth manifold of dimension $n \geq 2$ with a conformal class of metrics of signature (p,q). Let M_+ be a manifold with boundary satisfying $\partial M_+ = M$. Let r denote a defining function for ∂M_+; i.e., $r \in C^\infty(M_+)$ satisfies $r>0$ in the interior M_+°, $r=0$ on M, and $dr \neq 0$ on M. A smooth metric g_+ on M_+° of signature $(p+1,q)$ is said to be conformally compact if $r^2 g_+$ extends smoothly to M_+ and $r^2 g_+|_M$ is nondegenerate (so $r^2 g_+$ has signature $(p+1,q)$ also on M). A conformally compact metric is said to have conformal infinity $(M,[g])$ if $r^2 g_+|_{TM} \in [g]$. These conditions are independent of the choice of defining function r.

In the following, we will be concerned only with behavior near M. We will identify M_+ with an open neighborhood of $M \times \{0\}$ in $M \times [0,\infty)$, and r will denote the coordinate in the second factor. We will use lowercase Greek indices to label objects on M_+. Let $S_{\alpha\beta}$ be a symmetric 2-tensor field in an open neighborhood of $M \times \{0\}$ in $M \times [0,\infty)$. For $m \geq 0$, we will write $S = O^+_{\alpha\beta}(r^m)$ if $S = O(r^m)$ and $\operatorname{tr}_g(i^*(r^{-m}S)) = 0$ on M, where $i : M \to M \times [0,\infty)$ is $i(x) = (x,0)$ and g is a metric in the conformal class $[g]$.

Definition 4.1. A Poincaré metric for $(M,[g])$, where $[g]$ is a conformal class of signature (p,q) on M, is a conformally compact metric g_+ of signature $(p+1,q)$ on M_+°, where M_+ is an open neighborhood of $M\times\{0\}$ in $M\times[0,\infty)$, such that

(1) g_+ has conformal infinity $(M,[g])$.

(2) If n is odd or $n = 2$, then $\mathrm{Ric}(g_+) + ng_+$ vanishes to infinite order along M.

If $n \geq 4$ is even, then $\mathrm{Ric}(g_+) + ng_+ \in O^+_{\alpha\beta}(r^{n-2})$.

Alternatively, one can consider metrics g_- on M°_+ of signature $(p, q+1)$ such that $\mathrm{Ric}(g_-) - ng_-$ vanishes to the stated order. This is equivalent to the above upon taking $g_- = -g_+$, with $g \to -g$ and $(p,q) \to (q,p)$.

If g_+ is a conformally compact metric, then $|dr/r|_{g_+} = |dr|_{r^2 g_+}$ extends smoothly to M_+. The conformal transformation law for the curvature tensor shows that all sectional curvatures of g_+ approach $-|dr/r|^2_{g_+}$ at a boundary point (see [M]). We will say that a conformally compact metric g_+ is asymptotically hyperbolic if $|dr/r|_{g_+} = 1$ on M. A Poincaré metric is asymptotically hyperbolic.

There is a normal form for asymptotically hyperbolic metrics analogous to the normal form for pre-ambient metrics discussed in Chapter 2.

Definition 4.2. An asymptotically hyperbolic metric g_+ is said to be in *normal form* relative to a metric g in the conformal class if $g_+ = r^{-2}\left(dr^2 + g_r\right)$, where g_r is a 1-parameter family of metrics on M of signature (p,q) such that $g_0 = g$.

Proposition 4.3. *Let g_+ be an asymptotically hyperbolic metric on M°_+ and let g be a metric in the conformal class. Then there exists an open neighborhood $\mathcal{U}$ of $M \times \{0\}$ in $M \times [0,\infty)$ on which there is a unique diffeomorphism ϕ from $\mathcal{U}$ into M_+ such that $\phi|_M$ is the identity map, and such that $\phi^* g_+$ is in normal form relative to g on $\mathcal{U}$.*

We refer to §5 of [GrL] for the proof. The proof in [GrL] is for the case $M = S^n$ and g_+ positive definite, but the same argument applies in the general case, arguing as in the proof of Proposition 2.8 if M is noncompact.

We will say that an asymptotically hyperbolic metric g_+ on M°_+ is *even* if $r^2 g_+$ is the restriction to M_+ of a smooth metric h on an open set $\mathcal{V} \subset M \times (-\infty, \infty)$ containing M_+, such that $\mathcal{V}$ and h are invariant under $r \to -r$. We will say that a diffeomorphism ψ from M_+ into $M \times [0,\infty)$ satisfying $\psi|_{M\times\{0\}} = Id$ is even if ψ is the restriction of a diffeomorphism of such an open set $\mathcal{V}$ which commutes with $r \to -r$. If ψ is an even diffeomorphism and g_+ is an even asymptotically hyperbolic metric, then $\psi^* g_+$ is also even. An examination of the proof in [GrL] shows that if g_+ in Proposition 4.3 is even, then ϕ is also even.

The first main results of this chapter are the following analogues of Theorems 2.3 and 2.9.

Theorem 4.4. *Let $(M,[g])$ be a smooth manifold of dimension $n \geq 2$, equipped with a conformal class. Then there exists an even Poincaré metric for $(M,[g])$. Moreover, if g^1_+ and g^2_+ are two even Poincaré metrics* *for* $(M,[g])$ *defined on* $(M^1_+)^\circ$, $(M^2_+)^\circ$, *resp., then there are open subsets* $\mathcal{U}^1 \subset M^1_+$ *and* $\mathcal{U}^2 \subset M^2_+$ *containing* $M \times \{0\}$ *and an even diffeomorphism* $\phi : \mathcal{U}^1 \to \mathcal{U}^2$ *such that* $\phi|_{M\times\{0\}}$ *is the identity map, and such that*

(a) If $n = \dim M$ *is odd, then* $g^1_+ - \phi^* g^2_+$ *vanishes to infinite order at every point of* $M \times \{0\}$.

(b) If $n = \dim M$ *is even, then* $g^1_+ - \phi^* g^2_+ = O^+_{\alpha\beta}(r^{n-2})$.

Theorem 4.5. *Let* M *be a smooth manifold of dimension* $n \geq 2$ *and* g *a smooth metric on* M.

(A) There exists an even Poincaré metric g_+ *for* $(M,[g])$ *which is in normal form relative to* g.

(B) Suppose that g^1_+ *and* g^2_+ *are even Poincaré metrics for* $(M,[g])$, *both of which are in normal form relative to* g. *If* n *is odd, then* $g^1_+ - g^2_+$ *vanishes to infinite order at every point of* $M \times \{0\}$. *If* n *is even, then* $g^1_+ - g^2_+ = O^+_{\alpha\beta}(r^{n-2})$.

Theorem 4.4 follows from Theorem 4.5 and Proposition 4.3 just as in the proof of Theorem 2.3. Theorem 4.5 will be proven as a consequence of Theorem 2.9 after we establish the equivalence of straight ambient metrics and even Poincaré metrics in normal form.

Let $(\widetilde{\mathcal{G}},\widetilde{g})$ be a straight pre-ambient space for $(M,[g])$. It follows from Proposition 2.4 that $\|T\|^2$ vanishes exactly to first order on $\mathcal{G} \times \{0\} \subset \widetilde{\mathcal{G}}$. Therefore (shrinking $\widetilde{\mathcal{G}}$ if necessary) the hypersurface $\mathcal{H} = \widetilde{\mathcal{G}} \cap \{\|T\|^2 = -1\}$ lies on one side of $\mathcal{G} \times \{0\}$. Since $\|T\|^2$ is homogeneous of degree 2 with respect to the dilations, it follows (shrinking $\widetilde{\mathcal{G}}$ again if necessary) that each dilation orbit in $\widetilde{\mathcal{G}}$ on this side intersects $\mathcal{H}$ exactly once. We extend the projection $\pi : \mathcal{G} \to M$ to $\pi : \widetilde{\mathcal{G}} \subset \mathcal{G} \times \mathbb{R} \to M \times \mathbb{R}$ by acting in the first factor. Define $\chi : M \times \mathbb{R} \to M \times [0,\infty)$ by $\chi(x,\rho) = \left(x, \sqrt{2|\rho|}\right)$. Then (shrinking $\widetilde{\mathcal{G}}$ yet again if necessary) there is an open set M_+ in $M \times [0,\infty)$ containing $M \times \{0\}$ so that $\chi \circ \pi|_{\mathcal{H}} : \mathcal{H} \to M^\circ_+$ is a diffeomorphism. In the following, we allow ourselves to shrink the domains of definition of $\widetilde{g}$ and g_+ without further mention.

Proposition 4.6. *If* $(\widetilde{\mathcal{G}},\widetilde{g})$ *is a straight pre-ambient space for* $(M,[g])$ *and* $\mathcal{H}$ *and* M_+ *are as above, then*

$$g_+ := \left((\chi \circ \pi|_{\mathcal{H}})^{-1}\right)^* \widetilde{g} \tag{4.1}$$

is an even asymptotically hyperbolic metric with conformal infinity $(M,[g])$. *If* $\widetilde{g}$ *is in normal form relative to a metric* $g \in [g]$, *then* g_+ *is also in normal form relative to* g. *Every even asymptotically hyperbolic metric* g_+ *with conformal infinity* $(M,[g])$ *is of the form* (4.1) *for some straight pre-ambient metric* $\widetilde{g}$ *for* $(M,[g])$*. If* g_+ *is in normal form relative to* g*, then* $\widetilde{g}$ *can be taken to be in normal form relative to* g*, and in this case* $\widetilde{g}$ *on* $\{\|T\|^2 \leq 0\}$ *is uniquely determined by* g_+*.*

Proof. Choose a metric g in the conformal class at infinity, with corresponding identification $\mathcal{G} \times \mathbb{R} \cong \mathbb{R}_+ \times M \times \mathbb{R}$. Set $\widetilde{g}_{00} = a$, so $a = a(x,\rho)$ is smooth and homogeneous of degree 0 with respect to t. Then $\|T\|^2 = at^2$. According to condition (2) of Proposition 2.4, $\widetilde{g}$ is straight if and only if it has the form

$$\widetilde{g} = adt^2 + tdtda + t^2h, \tag{4.2}$$

where $h = h(x,\rho,dx,d\rho)$ is a smooth quadratic form defined in a neighborhood of $M \times \{0\}$ in $M \times \mathbb{R}$. The initial condition $\iota^*\widetilde{g} = \mathbf{g}_0$ is equivalent to $a(x,0) = 0$ and $h(x,0,dx,0) = g(x,dx)$, and nondegeneracy of $\widetilde{g}$ is equivalent to $\partial_\rho a \neq 0$ when $\rho = 0$.

Now $\mathcal{H} = \{at^2 = -1\}$. On $\{a < 0\}$, introduce new variables $u > 0$ and $s > 0$ by $a = -u^2$, $s = ut$. Elementary computation shows that $adt^2 + tdtda = s^2u^{-2}du^2 - ds^2$. Therefore, in terms of these variables $\widetilde{g}$ can be written

$$\widetilde{g} = s^2u^{-2}(h + du^2) - ds^2.$$

This is the cone metric over the base $u^{-2}(h + du^2)$. Thus every straight pre-ambient metric is a cone metric of this form. In the new variables, $\mathcal{H}$ is defined by the equation $s = 1$. So the restriction of $\widetilde{g}$ to $T\mathcal{H}$ is the metric $u^{-2}(h + du^2)$.

First suppose that $\widetilde{g}$ is in normal form relative to g. Comparing Lemma 3.1 with (4.2), one sees that this is equivalent to the conditions that $\partial_\rho a = 2$ and $h = g_\rho(x,dx)$, where g_ρ is a smooth 1-parameter family of metrics on M satisfying $g_0 = g$. We obtain $a = 2\rho$ and so $u = \sqrt{-2\rho}$. By the definition of χ, we see that $u = r$ is the coordinate in the second factor of $M \times [0,\infty)$. Hence g_+ defined by (4.1) is just the metric

$$g_+ = r^{-2}\left(dr^2 + g_{-\frac{1}{2}r^2}\right) \tag{4.3}$$

on $M \times [0,\infty)$. Clearly g_+ is an even asymptotically hyperbolic metric with conformal infinity $(M,[g])$ in normal form relative to g.

In the general case, we have $g_+ = u^{-2}(h + du^2)$ with $u = \sqrt{-a}$ and $h = h(x,\rho,dx,d\rho)$. Since r is given by $r = \sqrt{2|\rho|}$ and a vanishes exactly to first order at $\rho = 0$, we can write $u = rb(x,r^2)$ for a positive smooth function b. Using this and writing $\rho = \pm\frac{1}{2}r^2$, one sees easily that g_+ is an even asymptotically hyperbolic metric with conformal infinity $(M,[g])$.

To see that every even asymptotically hyperbolic metric with conformal infinity $(M,[g])$ is of this form, take $\widetilde{g}$ to be given by (4.2) with $a = 2\rho$ so that $g_+ = r^{-2}(h + dr^2)$ with $\rho = -\frac{1}{2}r^2$. Writing

$$h = h_{ij}(x,\rho)dx^idx^j + 2h_{i\infty}(x,\rho)dx^id\rho + h_{\infty\infty}(x,\rho)d\rho^2$$

gives

$$h + dr^2 = h_{ij}(x, -\tfrac{1}{2}r^2)dx^i dx^j - 2rh_{i\infty}(x, -\tfrac{1}{2}r^2)dx^i dr \\ + \left[1 + r^2 h_{\infty\infty}(x, -\tfrac{1}{2}r^2)\right] dr^2.$$

An asymptotically hyperbolic metric g_+ is even if and only if the Taylor expansions of the $dx^i dx^j$ coefficients of $r^2 g_+$ have only even terms, the Taylor expansions of the $dx^i dr$ coefficients of $r^2 g_+$ have only odd terms, and the dr^2 coefficient of $r^2 g_+$ equals 1 at $r = 0$ and has Taylor expansion with only even terms. Clearly, any such g_+ can be written in the form $r^{-2}(h + dr^2)$ for some h with all $h_{\alpha\beta}(x, \rho)$ smooth.

In general, there are many straight pre-ambient metrics $\widetilde{g}$ such that (4.1) gives the same metric g_+: a given such $\widetilde{g}$ can be pulled back by a diffeomorphism of $\widetilde{\mathcal{G}}$ which covers the identity on $M \times \mathbb{R}$ but which smoothly rescales the $\mathbb{R}_+$-fibers by a factor depending on the point in the base (and which restricts to the identity on $\mathcal{G} \times \{0\}$). If, however, g_+ is in normal form and $\widetilde{g}$ is required also to be in normal form, then the determination $a = 2\rho$ is forced. In this case, we saw above that if h in (4.2) is written $h = g_\rho(x, dx)$, then g_+ is given by (4.3). Clearly, g_+ uniquely determines h on $\{\rho \leq 0\}$. □

We summarize the relation for metrics in normal form. A straight pre-ambient metric in normal form can be written

$$\widetilde{g} = 2\rho dt^2 + 2tdtd\rho + t^2 g_\rho \tag{4.4}$$

for a 1-parameter family of metrics g_ρ on M. Under the change of variables $-2\rho = r^2$, $s = rt$ on $\{\rho \leq 0\}$, $\widetilde{g}$ takes the form

$$\widetilde{g} = s^2 g_+ - ds^2 \tag{4.5}$$

where g_+ is given by (4.3).

We remark that the hypothesis in Proposition 4.6 that $\widetilde{g}$ is straight is important: if $\widetilde{g}$ is not assumed to be straight, then the metric g_+ defined by (4.1) need not be asymptotically hyperbolic. Also, a cone metric $s^2 k - ds^2$ over any base (N, k), where N is a manifold and k a metric on N, is straight in the sense that for each $s > 0$, $p \in N$, the curve $\lambda \to (\lambda s, p)$ is a geodesic.

The relation between the curvature of a cone metric and that of the base is well-known. We include the derivation for completeness.

Proposition 4.7. *Let g_+ be a metric on a manifold M_+ of dimension $n+1$, and define a metric $\widetilde{g} = s^2 g_+ - ds^2$ on $M_+ \times \mathbb{R}_+$. Then*

$$\begin{aligned} \mathrm{Rm}(\widetilde{g}) &= s^2 \left[\mathrm{Rm}(g_+) + g_+ \owedge g_+\right] \\ \mathrm{Ric}(\widetilde{g}) &= \mathrm{Ric}(g_+) + n g_+ \\ R(\widetilde{g}) &= s^{-2} \left[R(g_+) + n(n+1)\right] \end{aligned}$$

Here Rm *denotes the Riemann tensor viewed as a covariant 4-tensor, and for symmetric 2-tensors u, v we write*

$$2(u \owedge v)_{IJKL} = u_{IK}v_{JL} - u_{IL}v_{JK} + u_{JL}v_{IK} - u_{JK}v_{IL},$$

so that $g_+ \owedge g_+$ is the curvature tensor of constant sectional curvature $+1$. Also, tensors on M_+ are implicitly pulled back to $M_+ \times \mathbb{R}_+$.

Proof. Set $s = e^y$; then $\widetilde{g} = e^{2y}\left(g_+ - dy^2\right)$ is a conformal multiple of a product metric. Under a conformal change $\widetilde{g} = e^{2y}h$, the curvature tensor transforms by

$$\operatorname{Rm}(\widetilde{g}) = e^{2y}\left[\operatorname{Rm}(h) + 2\Lambda \owedge h\right],$$

where $\Lambda = -\nabla_h^2 y + dy^2 - \frac{1}{2}|dy|_h^2\, h$. For $h = g_+ - dy^2$ we have $\operatorname{Rm}(h) = \operatorname{Rm}(g_+)$, $\nabla_h^2 y = 0$ and $|dy|_h^2 = -1$ so that $\Lambda = \frac{1}{2}(g_+ + dy^2)$. Thus $2\Lambda \owedge h = (g_+ + dy^2) \owedge (g_+ - dy^2) = g_+ \owedge g_+$, which gives the first equation. The second and third equations follow by contraction. $\square$

Proposition 4.7 shows in particular that $\widetilde{g}$ is flat if and only if g_+ has constant curvature -1, $\widetilde{g}$ is Ricci-flat if and only if $\operatorname{Ric}(g_+) = -ng_+$, and $\widetilde{g}$ is scalar-flat if and only if g_+ has constant scalar curvature $-n(n+1)$. Also, it is immediate from the equation for $\operatorname{Rm}(\widetilde{g})$ in Proposition 4.7 that $\partial_s \lrcorner \operatorname{Rm}(\widetilde{g}) = 0$; compare with Lemma 3.2 and the discussion before (6.1).

Proof of Theorem 4.5. Given M and g, Theorem 2.9 shows that there exists an ambient metric in normal form relative to g, which by Propositions 3.3 and 3.4 can be taken to be straight. According to Proposition 4.6, the metric g_+ defined by (4.1) is even and in normal form relative to g, and (4.5) holds. Proposition 4.7 shows that the Ricci curvature of g_+ is constant to the correct order, so that g_+ is a Poincaré metric for $(M, [g])$. This proves (A). Part (B) follows similarly from part (B) of Theorem 2.9: every even Poincaré metric g_+ in normal form gives rise to a straight ambient metric in normal form by (4.5). $\square$

Thus for metrics in normal form, the formal asymptotics for Poincaré metrics is entirely equivalent to that for straight ambient metrics under the change of variable $\rho = -\frac{1}{2}r^2$. In particular, via this change of variable one can easily write down the analogues for g_+ of (3.17), (3.18), (3.19), (3.21), and Proposition 3.5. For future reference, we observe from (3.6) that for $n \geq 3$, the expansion for a Poincaré metric $g_+ = r^{-2}\left(dr^2 + g_r\right)$ in normal form begins

$$(g_r)_{ij} = g_{ij} - P_{ij}r^2 + \dots \tag{4.6}$$

(We have returned to our original notation of g_r for the M-component of $r^2 g_+$.)

There is a substantial literature concerning the asymptotics of Poincaré metrics. See for example [GrH1] and [HSS] for direct analyses of the asymptotics up to order n in r from the Poincaré metric point of view, and [R] for a study of the asymptotics in the context of general relativity.

As discussed in Chapter 3, Theorems 3.7, 3.9 and 3.10 describe all formal expansions involving powers of ρ and $\log|\rho|$ of generalized ambient metrics in normal form. Proposition 4.6 extends with the same proof to the case of metrics with such expansions; the relations between the expansions and regularity of $\widetilde{g}$ and g_+ are determined by the change of variable $\rho = -\frac{1}{2}r^2$. This gives the following result describing all formal expansions $g_+ = r^{-2}\left(dr^2 + g_r\right)$ solving $\mathrm{Ric}(g_+) = -ng_+$.

Theorem 4.8. *Let M be a smooth manifold of dimension n, and let g be a smooth metric of signature (p,q) and h a smooth symmetric 2-tensor on M such that $g^{ij}h_{ij} = 0$.*

- *If $n = 2$ and if h satisfies $h_{ij,}{}^{j} = -\frac{1}{2}R_{,i}$, then there exists an even Poincaré metric in normal form relative to g, such that* $\mathrm{tf}\left(\partial_r^2 g_r|_{r=0}\right) = h$. *These conditions uniquely determine g_r to infinite order at $r = 0$. (The solution g_r can be written explicitly; see Chapter 7.)*
- *If $n \geq 3$ is odd and if h satisfies $h_{ij,}{}^{j} = 0$, then there exists a Poincaré metric in normal form relative to g, such that* $\mathrm{tf}\left(\partial_r^n g_r|_{r=0}\right) = h$. *These conditions uniquely determine g_r to infinite order at $r = 0$, and the solution satisfies $\mathrm{tr}_g\left(\partial_r^n g_r|_{r=0}\right) = 0$.*
- *Let $n \geq 4$ be even. There is a natural pseudo-Riemannian invariant 1-form $\overline{D}_i$ so that if h satisfies $h_{ij,}{}^{j} = \overline{D}_i$, then there exists a metric $g_+ = r^{-2}\left(dr^2 + g_r\right)$ satisfying $g_0 = g$ and* $\mathrm{Ric}(g_+) = -ng_+$ *to infinite order, such that g_r has an expansion of the form*

$$g_r \sim \sum_{N=0}^{\infty} g_r^{(N)} \left(r^n \log r\right)^N,$$

where each of the $g_r^{(N)}$ is a smooth family of symmetric 2-tensors on M even in r, and $\mathrm{tf}\left(\partial_r^n g_r^{(0)}\big|_{r=0}\right) = h$. *These conditions uniquely determine the $g_r^{(N)}$ to infinite order at $r = 0$. The solution g_r is smooth (i.e., $g_r^{(N)}$ vanishes to infinite order for $N \geq 1$) if and only if the obstruction tensor $\mathcal{O}_{ij}$ vanishes on M.*

The 1-form $\overline{D}_i$ is given by

$$\overline{D}_i = \frac{(-2)^{-n/2} n!}{(n/2)!} D_i,$$

where D_i is the 1-form appearing in Theorem 3.10.

As for ambient metrics, if g and h are real-analytic, then the formal series for g_r converges.

For n odd, the even solution given by Theorem 4.5 is the one determined by Theorem 4.8 upon taking $h = 0$.

We close this chapter by describing an alternate interpretation of Poincaré metrics in terms of projective geometry. This is motivated by the Klein model of hyperbolic space, which is the metric

$$(1-|x|^2)^{-1}\sum(dx^i)^2 + (1-|x|^2)^{-2}\left(\sum x^i dx^i\right)^2$$

on the ball, whose geodesics are straight lines. Begin by recalling that two torsion free connections ∇ and ∇' on the tangent bundle of a manifold are said to be projectively equivalent if they have the same geodesics up to parametrization, and that this condition is equivalent to the requirement that their difference tensor has the form $v_{(i}\delta_{j)}{}^k$ for some 1-form v on M. Given a manifold with boundary N, consider the class of metrics on N° which near the boundary have the form $h/\rho + d\rho^2/4\rho^2$, where ρ is a defining function for the boundary and h is a symmetric 2-tensor which is smooth up to the boundary and for which $h|_{T\partial N}$ is nondegenerate of signature (p,q). It is easily checked that this class of metrics is independent of the choice of defining function so is invariantly associated to N. It is also easily checked that the conformal class of $h|_{T\partial N}$ is also independent of the choice of ρ, so should be called the conformal infinity by analogy with the conformally compact case. Elementary calculations show that if ∇ denotes the Levi-Civita connection of such a metric, then there is a connection ∇' which is smooth up to the boundary so that the difference tensor $\nabla - \nabla'$ takes the form $v_{(i}\delta_{j)}{}^k$ where $v = -d\rho/\rho$. (The equivalence class of v modulo smooth 1-forms is independent of ρ.) Thus it makes sense to call such metrics projectively compact; their geodesics are the same as those of a smooth connection up to parametrization. In the projective formulation, a Poincaré metric is then defined to be a projectively compact metric with prescribed conformal infinity and with constant Ricci curvature $-n$. The change of variable $\rho = r^2$ transforms the class of projectively compact metrics to the class of even asymptotically hyperbolic metrics. As the construction of Proposition 4.6 shows, the projectively compact metrics are more directly related to pre-ambient metrics than are the asymptotically hyperbolic metrics. A defining function ρ and the smooth structure on the space where the projectively compact metrics live are induced directly from that on the ambient space without the introduction of the square root.

Chapter Five

Self-dual Poincaré Metrics

In [LeB], LeBrun showed using twistor methods that if g is a real-analytic metric on an oriented real-analytic 3-manifold M, then $[g]$ is the conformal infinity of a real-analytic self-dual Einstein metric on a deleted collar neighborhood of $M \times \{0\}$ in $M \times [0, \infty)$, uniquely determined up to real-analytic diffeomorphism. As mentioned in [FG], LeBrun's result can be proved as an application of our formal theory of Poincaré metrics. In this chapter we show that the corresponding formal power series statement is a consequence of Theorem 4.8. The self-duality condition can be viewed as providing a conformally invariant specification of the formally undetermined term $\partial_r^3 g_r|_{r=0}$.

Let M be an oriented 3-manifold. Give $M \times [0, \infty)$ the induced orientation determined by the requirement that $(\partial_r, e_1, e_2, e_3)$ is positively oriented for a positively oriented frame (e_1, e_2, e_3) for T_pM, $p \in M$. Here, as above, r denotes the coordinate in the $[0, \infty)$ factor. Throughout this chapter we will use lowercase Greek indices to label objects on $M \times [0, \infty)$, lowercase Latin indices for objects on M, and a '0' index for the $[0, \infty)$ factor.

Let g be a metric on M of signature (p, q) and let $g_+ = r^{-2}\left(dr^2 + g_r\right)$ be an asymptotically hyperbolic metric on M_+° in normal form relative to g, where M_+ is an open neighborhood of $M \times \{0\}$ in $M \times [0, \infty)$. Set $\overline{g} = dr^2 + g_r$. We denote by μ the volume form of g_r on M (the dependence on r is to be understood), and by $\overline{\mu}$ the volume form of $\overline{g}$. In terms of a positively oriented coordinate system on M, these are given by $\mu_{ijk} = \overline{\mu}_{0ijk} = \sqrt{|\det g_r|}\,\epsilon_{ijk}$, where ϵ_{ijk} denotes the sign of the permutation. We use $\overline{g}$ to raise and lower Greek indices and g_r to raise and lower Latin indices.

The metric $\overline{g}$ and the orientation determine a Hodge $*$ operator on $\Lambda^2(M_+)$ which is conformally invariant so agrees with the $*$ operator for g_+. This induces an operator $* : \mathcal{W} \to \mathcal{W}$, where $\mathcal{W} \subset \otimes^4 T^* M_+$ denotes the bundle of algebraic Weyl tensors; i.e., 4-tensors with curvature tensor symmetry which are trace-free with respect to $\overline{g}$ (or equivalently g_+). If $W_{\alpha\beta\gamma\delta}$ is a section of $\mathcal{W}$, then

$$(*W)_{\alpha\beta\gamma\delta} = \tfrac{1}{2}\,\overline{\mu}_{\alpha\beta}{}^{\rho\sigma} W_{\rho\sigma\gamma\delta}. \tag{5.1}$$

One has $*^2 = (-1)^q$. We assume throughout the rest of this chapter that q is even (i.e., $q = 0$ or $q = 2$) so that $*^2 = 1$. Then $\mathcal{W} = \mathcal{W}^+ \oplus \mathcal{W}^-$, where $\mathcal{W}^\pm$ denote the ± 1 eigenspaces of $*$.

We will use an alternate description of the bundles $\mathcal{W}^{\pm}$. For $(p,r) \in M_+ \subset M \times \mathbb{R}$, define a map $s : \mathcal{W}_{(p,r)} \to \odot^2 T^*_p M$ by

$$s(W)_{ij} = W_{0i0j}.$$

Observe that $s(W)$ is trace-free with respect to g_r since W is trace-free with respect to $\overline{g}$:

$$0 = \overline{g}^{\alpha\beta} W_{0\alpha 0\beta} = (g_r)^{ij} W_{0i0j} + W_{0000} = (g_r)^{ij} s(W)_{ij}.$$

We denote by $\odot^2_0 T^*M$ the bundle on M_+ whose fiber at (p,r) consists of the symmetric 2-tensors in the M factor which are trace-free with respect to $g_r(p)$. No confusion should arise with this notation since this bundle restricts to $M \subset M_+$ to the bundle for which this notation is usually used, and the context will make clear whether the bundle is considered as defined on M or M_+.

Lemma 5.1. $s|_{\mathcal{W}^{\pm}} : \mathcal{W}^{\pm} \to \odot^2_0 T^*M$ *is a bundle isomorphism. (Here s is restricted to either the bundle $\mathcal{W}^+$ or $\mathcal{W}^-$.)*

Proof. Since the bundles $\mathcal{W}^{\pm}$ and $\odot^2_0 T^*M$ all have rank 5, it suffices to show that $s|_{\mathcal{W}^{\pm}}$ is injective. If $W \in \mathcal{W}$, then

$$0 = \overline{g}^{\alpha\beta} W_{\alpha i \beta j} = (g_r)^{kl} W_{kilj} + W_{0i0j}.$$

So if $s(W) = 0$, then $(g_r)^{kl} W_{kilj} = 0$. But then W_{kilj} is a trace-free tensor in 3 dimensions with curvature tensor symmetry, so $W_{kilj} = 0$.

If also $W \in \mathcal{W}^{\pm}$, then (5.1) gives

$$W_{0jkl} = \pm \tfrac{1}{2} \overline{\mu}_{0j}{}^{\rho\sigma} W_{\rho\sigma kl}.$$

The terms all vanish for which ρ or σ is 0, and the terms vanish in which neither ρ nor σ is 0 since $W_{ijkl} = 0$. Thus all components $W_{0jkl} = 0$. This accounts for all possible nonzero components of W, so $W = 0$ and $s|_{\mathcal{W}^{\pm}}$ is injective. □

We use Lemma 5.1 to represent sections of $\mathcal{W}^{\pm}$ as one-parameter families of tensors on M. Let $\overline{W}$ denote the Weyl tensor of $\overline{g}$ and $\overline{W}^{\pm}$ its $\pm$ self-dual parts. Then for either choice of $\pm$, $s(\overline{W}^{\pm})$ can be regarded as a one-parameter family parametrized by r of sections of the fixed bundle $\odot^2 T^*M$ on M. Thus it makes sense to differentiate this family with respect to r. The following proposition gives an expression for the first derivative of $s(\overline{W}^{\pm})$ in terms of derivatives of g_r. We use $'$ to denote differentiation with respect to r and we sometimes suppress the subscript r on g_r.

Proposition 5.2. *Let g_r be a one-parameter family of metrics on M satisfying $g_r' = 0$ at $r = 0$, and let $\overline{g} = dr^2 + g_r$. Then at $r = 0$ we have*

$$s(\overline{W}^{\pm})'_{ij} = -\tfrac{1}{8}\, \mathrm{tf}(g'''_{ij}) \pm \tfrac{1}{4} \nabla_l g''_{k(i} \mu_{j)}{}^{kl}.$$

On the right-hand side, tf, ∇ *and* μ *are with respect to the initial metric* g_0.

We remark that the star operator on 2-forms on M with respect to g is given by $(*\eta)_i = \frac{1}{2}\mu_i{}^{kl}\eta_{kl}$. Thus the last term above can be interpreted as a multiple of $\mathrm{Sym}\,*dg''$, where d acts on g'' as a 1-form-valued 1-form, $*$ acts on the 2-form indices generated by d, and Sym symmetrizes over the two indices of $*dg''$.

Proof. Specializing the indices in

$$\overline{W}_{\alpha\beta\gamma\delta} = \overline{R}_{\alpha\beta\gamma\delta} - \left(\overline{P}_{\alpha\gamma}\overline{g}_{\beta\delta} - \overline{P}_{\alpha\delta}\overline{g}_{\beta\gamma} - \overline{P}_{\beta\gamma}\overline{g}_{\alpha\delta} + \overline{P}_{\beta\delta}\overline{g}_{\alpha\gamma}\right) \tag{5.2}$$

and using the form of $\overline{g}$ gives

$$\overline{W}_{0i0j} = \overline{R}_{0i0j} - \left(\overline{P}_{00}g_{ij} + \overline{P}_{ij}\right).$$

But we know that the left-hand side is trace-free in ij with respect to g_r, so taking the trace-free part with respect to g_r and recalling the definition of $\overline{P}$ gives

$$\overline{W}_{0i0j} = \mathrm{tf}\left(\overline{R}_{0i0j} - \tfrac{1}{2}\overline{R}_{ij}\right).$$

Now $\overline{R}_{ij} = \overline{g}^{\alpha\beta}\overline{R}_{\alpha i\beta j} = \overline{R}_{0i0j} + g^{kl}\overline{R}_{ikjl}$, giving

$$\overline{W}_{0i0j} = \tfrac{1}{2}\,\mathrm{tf}\left(\overline{R}_{0i0j} - g^{kl}\overline{R}_{ikjl}\right). \tag{5.3}$$

Next, (5.1) gives $(*\overline{W})_{0i0j} = \frac{1}{2}\overline{\mu}_{0i}{}^{\rho\sigma}\overline{W}_{\rho\sigma 0j} = \frac{1}{2}\overline{\mu}_{0i}{}^{kl}\overline{W}_{kl0j}$. Specializing (5.2) again, substituting, and simplifying gives easily

$$(*\overline{W})_{0i0j} = \tfrac{1}{2}\overline{\mu}_{0i}{}^{kl}\left(\overline{R}_{0jkl} - \overline{R}_{0k}g_{jl}\right).$$

However, $\overline{\mu}_{0i}{}^{kl}\overline{R}_{0k}g_{jl} = \overline{\mu}_{0i}{}^{k}{}_{j}\overline{R}_{0k}$ is skew in ij and $(*\overline{W})_{0i0j}$ is symmetric in ij, so we can symmetrize to obtain

$$(*\overline{W})_{0i0j} = \tfrac{1}{2}\overline{\mu}_{0(i}{}^{kl}\overline{R}_{j)0lk}. \tag{5.4}$$

Now it is straightforward to calculate the curvature tensor of $\overline{g} = dr^2 + g_r$; one obtains the curvature tensor of a product metric plus extra terms involving r-derivatives of g_r. The result (cf. Gauss-Codazzi equations) is

$$\begin{aligned}
\overline{R}_{ijkl} &= R_{ijkl} + \tfrac{1}{4}\left(g'_{il}g'_{jk} - g'_{ik}g'_{jl}\right),\\
\overline{R}_{0jkl} &= \tfrac{1}{2}\left(\nabla_l g'_{jk} - \nabla_k g'_{jl}\right),\\
\overline{R}_{0i0j} &= -\tfrac{1}{2}\,g''_{ij} + \tfrac{1}{4}\,g^{rs}g'_{ir}g'_{js},
\end{aligned}$$

where R_{ijkl} and ∇ are with respect to g_r as a metric on M with r fixed. Substituting into (5.3) and (5.4) gives

$$\begin{aligned}
\overline{W}_{0i0j} &= -\tfrac{1}{4}\,\mathrm{tf}\left(g''_{ij} + 2R_{ij} - \tfrac{1}{2}g^{kl}g'_{kl}g'_{ij}\right),\\
(*\overline{W})_{0i0j} &= \tfrac{1}{2}\,\nabla_l g'_{k(i}\mu_{j)}{}^{kl}.
\end{aligned}$$

Now differentiate these equations with respect to r at $r = 0$. Since $g' = 0$ at $r = 0$, it follows that raising indices, taking the trace-free part and applying

∇ all commute with the differentiation. Similarly, the derivatives of R_{ij} and μ_{jkl} vanish at $r = 0$. Thus we have at $r = 0$

$$\overline{W}'_{0i0j} = -\tfrac{1}{4}\operatorname{tf}\left(g'''_{ij}\right),$$
$$(*\overline{W})'_{0i0j} = \tfrac{1}{2}\nabla_l g''_{k(i}\mu_{j)}{}^{kl}.$$

Adding and subtracting yields the stated formula. □

On an oriented pseudo-Riemannian 3-manifold of (arbitrary) signature (p, q), the Cotton tensor C_{jkl} can be reinterpreted as a trace-free symmetric 2-tensor. Define

$$\mathcal{C}_{ij} = \mu_i{}^{kl}C_{jkl}.$$

We sometimes write $\mathcal{C}(g)$ to indicate the underlying metric. The fact that $C_{[jkl]} = 0$ implies that $\mathcal{C}_{ij}$ is trace-free:

$$g^{ij}\mathcal{C}_{ij} = \mu^{jkl}C_{jkl} = 0,$$

and the fact that C_{jkl} is trace-free implies that $\mathcal{C}_{ij}$ is symmetric:

$$\mu^{ijm}\mathcal{C}_{ij} = \mu^{ijm}\mu_i{}^{kl}C_{jkl} = (-1)^q\left(g^{jk}g^{ml} - g^{jl}g^{mk}\right)C_{jkl} = 0.$$

The fact that $C_{jkl,}{}^j = 0$ implies that $\mathcal{C}_{ij,}{}^j = 0$.

We will say that a Poincaré metric g_+ is a $\pm$ self-dual Poincaré metric if $W^{\mp}(g_+)$ vanishes to infinite order. The main result of this chapter is the following theorem.

Theorem 5.3. *Let g be a smooth metric of signature (p, q) on a manifold M of dimension 3, with q even.*

(1) A Poincaré metric $g_+ = r^{-2}\left(dr^2 + g_r\right)$ in normal form relative to g is a $\pm$ self-dual Poincaré metric if and only if

$$\left(\partial_r^3 g_r\right)|_{r=0} = \pm 2\mathcal{C}(g). \tag{5.5}$$

(2) There exists a $\pm$ self-dual Poincaré metric g_+ in normal form relative to g, and such a g_+ is uniquely determined to infinite order.

Proof. If g_+ is a Poincaré metric in normal form relative to g, then Theorem 4.8 shows that $g'''_r|_{r=0}$ is trace-free and (4.6) shows that $g''_r|_{r=0} = -2P_{ij}$. Thus Proposition 5.2 shows that at $r = 0$ we have

$$\begin{aligned} s(\overline{W}^{\mp})'_{ij} &= -\tfrac{1}{8}g'''_{ij} \pm \tfrac{1}{2}\nabla_l P_{k(i}\mu_{j)}{}^{kl} \\ &= -\tfrac{1}{8}g'''_{ij} \pm \tfrac{1}{4}\mu_{(j}{}^{kl}C_{i)kl} \\ &= -\tfrac{1}{8}g'''_{ij} \pm \tfrac{1}{4}\mathcal{C}_{ij}. \end{aligned} \tag{5.6}$$

This must vanish if $W^{\mp}$ vanishes to infinite order, which establishes the necessity of (5.5).

For the converse, we must show that $\overline{W}^{\mp}$ vanishes to infinite order if (5.5) holds. For this we use the fact that $W^{\mp}(g_+)$ satisfies a first order equation as a consequence of the Bianchi identity and the fact that g_+ is Einstein. The contracted second Bianchi identity in dimension 4 can be written $\nabla^\alpha W_{\alpha\beta\gamma\delta} = -C_{\beta\gamma\delta}$. The Cotton tensor vanishes for Einstein metrics, so it follows that $\nabla^\alpha W_{\alpha\beta\gamma\delta} = O(r^\infty)$, where W, ∇ and index raising are with respect to g_+. Now rewrite $\nabla^\alpha W_{\alpha\beta\gamma\delta}$ in terms of $\overline{\nabla}$ and $\overline{W}$. The connections of g_+ and $\overline{g}$ are related by

$$\Gamma^\alpha_{\beta\gamma}(g_+) - \Gamma^\alpha_{\beta\gamma}(\overline{g}) = -r^{-1}\left(r_\beta\delta^\alpha{}_\gamma + r_\gamma\delta^\alpha{}_\beta - \overline{g}^{\alpha\lambda}r_\lambda\overline{g}_{\beta\gamma}\right).$$

Using this to transform the covariant derivative shows that

$$\nabla^\alpha W_{\alpha\beta\gamma\delta} = r^2\,\overline{\nabla}^\alpha W_{\alpha\beta\gamma\delta} + r r^\alpha W_{\alpha\beta\gamma\delta},$$

where on the right-hand side the index is raised using $\overline{g}$. Now substitute $W_{\alpha\beta\gamma\delta} = r^{-2}\overline{W}_{\alpha\beta\gamma\delta}$ to conclude that

$$r\,\overline{\nabla}^\alpha\overline{W}_{\alpha\beta\gamma\delta} - r^\alpha\overline{W}_{\alpha\beta\gamma\delta} = O(r^\infty).$$

Since $\overline{\nabla}$ commutes with $*$, this equation also holds for $\overline{W}^{\mp}$. Using $r^\alpha = \delta^\alpha{}_0$, we therefore obtain

$$r\,\overline{\nabla}^\alpha\overline{W}^{\mp}_{\alpha\beta\gamma\delta} - \overline{W}^{\mp}_{0\beta\gamma\delta} = O(r^\infty). \tag{5.7}$$

Suppose we know that $\overline{W}^{\mp} = O(r^s)$ for some s. Write $\overline{W}^{\mp} = r^sV$ and substitute into (5.7). One obtains $(s-1)r^sV_{0\beta\gamma\delta} + O(r^{s+1}) = O(r^\infty)$. So if $s \neq 1$, then $V_{0\beta\gamma\delta} = 0$ at $r = 0$. Taking $\beta\gamma\delta = i0j$ shows that $s(V) = 0$ at $r = 0$, which implies $V = 0$ at $r = 0$ by Lemma 5.1. Thus one concludes that $\overline{W}^{\mp} = O(r^{s+1})$ if $s \neq 1$.

We use this observation inductively. Begin with $s = 0$ to conclude that $\overline{W}^{\mp} = O(r)$ for any Poincaré metric in normal form. If also (5.5) holds, then (5.6) shows that $s(\overline{W}^{\mp})' = 0$ at $r = 0$, which gives $\overline{W}^{\mp} = O(r^2)$. Now the induction proceeds to all higher orders and one concludes that $\overline{W}^{\mp} = O(r^\infty)$ as desired.

Part (2) is an immediate consequence of (1) and Theorem 4.8. For any metric g on M, the tensor $\mathcal{C}(g)$ is trace-free and divergence-free. Take $h = \pm 2\mathcal{C}(g)$ in Theorem 4.8 to conclude the existence of a Poincaré metric in normal form relative to g which satisfies (5.5). Part (1) implies that this Poincaré metric is $\pm$ self-dual to infinite order. Uniqueness follows from the necessity of (5.5) in (1) together with the uniqueness statement in Theorem 4.8. □

It follows by putting metrics into normal form that two self-dual Poincaré metrics with the same conformal infinity agree up to diffeomorphism and up to terms vanishing to infinite order.

If g in Theorem 5.3 is real-analytic, then $\mathcal{C}(g)$ is also real-analytic, so the series for g_r converges as mentioned earlier. Thus one recovers LeBrun's result.

We remark that for definite signature, one can prove a version of part (1) of Theorem 5.3 concluding $\pm$ self-duality in an open set near the boundary for metrics which are Einstein in an open set and which satisfy (5.5). This is stated as Theorem 3 of [Gr2].

Chapter Six

Conformal Curvature Tensors

In this chapter we study conformal curvature tensors of a pseudo-Riemannian metric g. These are defined in terms of the covariant derivatives of the curvature tensor of an ambient metric in normal form relative to g. Their transformation laws under conformal change are given in terms of the action of a subgroup of the conformal group $O(p+1,q+1)$ on tensors. We assume throughout this chapter that $n \geq 3$.

Let g be a metric on a manifold M. By Theorem 2.9, there is an ambient metric in normal form relative to g, which by Proposition 2.6 we may take to be straight. Such a metric takes the form (3.14) on a neighborhood of $\mathbb{R}_+ \times M \times \{0\}$ in $\mathbb{R}_+ \times M \times \mathbb{R}$. Equations (3.17) determine the 1-parameter family of metrics $g_{ij}(x,\rho)$ on M in terms of the initial metric to infinite order for n odd and modulo $O(\rho^{n/2})$ for n even, except that also $g^{ij}\partial_\rho^{n/2} g_{ij}|_{\rho=0}$ is determined for n even. Each of the determined Taylor coefficients is a natural invariant of the initial metric g.

We consider the curvature tensor and its covariant derivatives for such an ambient metric $\widetilde{g}$. In general, we denote the curvature tensor of a pre-ambient metric by $\widetilde{R}$, with components $\widetilde{R}_{IJKL}$. Its r-th covariant derivative will be denoted $\widetilde{R}^{(r)}$, with components $\widetilde{R}^{(r)}_{IJKL,M_1\cdots M_r}$. Sometimes the superscript $^{(r)}$ will be omitted when the list of indices makes clear the value of r. Using (3.15), it is straightforward to calculate the curvature tensor of a metric of the form (3.14). One finds that $\widetilde{R}_{IJK0} = 0$ (another derivation of this is given in Proposition 6.1) and that the other components are given by

$$
\begin{aligned}
\widetilde{R}_{ijkl} &= t^2\left[R_{ijkl} + \frac{1}{2}(g_{il}g'_{jk} + g_{jk}g'_{il} - g_{ik}g'_{jl} - g_{jl}g'_{ik}) + \frac{\rho}{2}(g'_{ik}g'_{jl} - g'_{il}g'_{jk})\right]\\
\widetilde{R}_{\infty jkl} &= \frac{1}{2}t^2\left[\nabla_l g'_{jk} - \nabla_k g'_{jl}\right]\\
\widetilde{R}_{\infty jk\infty} &= \frac{1}{2}t^2\left[g''_{jk} - \frac{1}{2}g^{pq}g'_{jp}g'_{kq}\right].
\end{aligned}
\tag{6.1}
$$

Here $'$ denotes ∂_ρ, R_{ijkl} denotes the curvature tensor of the metric $g_{ij}(x,\rho)$ with ρ fixed, and ∇ its Levi-Civita connection. Components of the ambient covariant derivatives of curvature can then be calculated recursively starting with these formulae and using (3.16).

Fix a straight ambient metric $\widetilde{g}$ in normal form relative to g. We construct tensors on M from the covariant derivatives of curvature of $\widetilde{g}$ as follows. Choose an order $r \geq 0$ of covariant differentiation. Divide the set of symbols $IJKLM_1 \cdots M_r$ into three disjoint subsets labeled $\mathcal{S}_0$, $\mathcal{S}_M$ and $\mathcal{S}_\infty$. Set the indices in $\mathcal{S}_0$ equal to 0, those in $\mathcal{S}_\infty$ equal to ∞, and let those in $\mathcal{S}_M$ correspond to M in the decomposition $\mathbb{R}_+ \times M \times \mathbb{R}$. (In local coordinates, the indices in $\mathcal{S}_M$ vary between 1 and n.) Evaluate the resulting component $\widetilde{R}_{IJKL,M_1\cdots M_r}$ at $\rho = 0$ and $t = 1$. This defines a tensor on M which we denote by $\widetilde{R}^{(r)}_{\mathcal{S}_0,\mathcal{S}_M,\mathcal{S}_\infty}$. (Recall that the submanifold $\{\rho = 0,\ t = 1\}$ of $\mathbb{R}_+ \times M \times \mathbb{R}$ can be invariantly described as the image of g viewed as a section of $\mathcal{G}$.)

Consider for example the case $r = 0$. Since $\widetilde{R}_{IJK0} = 0$ as noted above, we must choose $\mathcal{S}_0 = \emptyset$ in order to get something nonzero. If we choose also $\mathcal{S}_\infty = \emptyset$ so that $\mathcal{S}_M = \{IJKL\}$, then the resulting tensor $\widetilde{R}^{(0)}_{\mathcal{S}_0,\mathcal{S}_M,\mathcal{S}_\infty}$ can be determined by setting $\rho = 0$ and $t = 1$ in the first line of (6.1). Recall from (3.6), (3.7) that $g'_{ij}|_{\rho=0} = 2P_{ij}$. This gives $\widetilde{R}_{ijkl}|_{\rho=0} = t^2 W_{ijkl}$, so in this case the tensor $\widetilde{R}^{(0)}_{\mathcal{S}_0,\mathcal{S}_M,\mathcal{S}_\infty}$ is the Weyl tensor of g. Similarly one finds that $\widetilde{R}_{\infty jkl}|_{\rho=0,\,t=1} = C_{jkl}$ is the Cotton tensor of g. The tensors for all other possibilities for the subsets $\mathcal{S}_0$, $\mathcal{S}_M$, $\mathcal{S}_\infty$ are determined from these and from $\widetilde{R}_{\infty jk\infty}$ by the usual symmetries of the curvature tensor. Now $\widetilde{R}_{\infty jk\infty}$ is given by the last line of (6.1). When $n = 4$, the trace-free part of $g''_{jk}|_{\rho=0}$ depends on the specific ambient metric which has been chosen and is not determined solely by g, so the same holds for $\widetilde{R}_{\infty jk\infty}$. However, when $n \neq 4$, one finds using the first line of (3.18) and the last line of (6.1) that $\widetilde{R}_{\infty jk\infty}|_{\rho=0,\,t=1} = -(n-4)^{-1}B_{jk}$. Thus the conformal curvature tensors which arise for $r = 0$ are precisely the Weyl, Cotton, and Bach tensors of g:

$$\widetilde{R}_{ijkl}|_{\rho=0,t=1} = W_{ijkl}, \quad \widetilde{R}_{\infty jkl}|_{\rho=0,\,t=1} = C_{jkl}, \quad \widetilde{R}_{\infty jk\infty}|_{\rho=0,\,t=1} = -\frac{B_{jk}}{n-4}. \tag{6.2}$$

Iterated covariant derivatives of $\widetilde{R}$ can be calculated recursively using (3.16). For example, one obtains

$$\widetilde{R}_{ijkl,m}|_{\rho=0,\,t=1} = V_{ijklm}, \qquad \widetilde{R}_{\infty jkl,m}|_{\rho=0,\,t=1} = Y_{jklm},$$

where

$$\begin{aligned} V_{ijklm} &= W_{ijkl,m} + g_{im}C_{jkl} - g_{jm}C_{ikl} + g_{km}C_{lij} - g_{lm}C_{kij} \\ Y_{jklm} &= C_{jkl,m} - P_m{}^i W_{ijkl} + (n-4)^{-1}(g_{km}B_{jl} - g_{lm}B_{jk}), \end{aligned} \tag{6.3}$$

and we assume that $n \neq 4$ for $\widetilde{R}_{\infty jkl,m}$.

The usual identities satisfied by covariant derivatives of curvature imply identities and relations amongst the conformal curvature tensors. For example, we conclude that

$$V_{ijklm} = V_{[ij][kl]m} \qquad V_{i[jkl]m} = 0 \qquad V_{ij[klm]} = 0$$
$$Y_{jklm} = Y_{j[kl]m} \qquad Y_{[jkl]m} = 0 \qquad Y_{j[klm]} = 0.$$

The differentiated Ricci identity for commuting ambient covariant derivatives gives relations involving the conformal curvature tensors. The asymptotic Ricci-flatness of the ambient metric gives relations amongst different conformal curvature tensors about which we will be more precise shortly. The covariant derivatives of ambient curvature also satisfy extra identities involving the infinitesimal dilation T arising from the homogeneity and straightness conditions, as follows.

Proposition 6.1. *The covariant derivatives of the curvature tensor of a straight pre-ambient metric* $\widetilde{g}$ *satisfy*

(1) $T^L \widetilde{R}_{IJKL,M_1\cdots M_r} = -\sum_{s=1}^r \widetilde{R}_{IJKM_s,M_1\cdots \widehat{M_s}\cdots M_r}$;

(2) $T^P \widetilde{R}_{IJKL,M_1\cdots M_s P M_{s+1}\cdots M_r} =$
$-(s+2)\widetilde{R}_{IJKL,M_1\cdots M_r} - \sum_{t=s+1}^r \widetilde{R}_{IJKL,M_1\cdots M_s M_t M_{s+1}\cdots \widehat{M_t}\cdots M_r}$.

Condition (1) in the case $r = 0$ is interpreted as the statement $T^L \widetilde{R}_{IJKL} = 0$, or equivalently $\widetilde{R}_{IJK0} = 0$, mentioned above. Note also that the case $s = r$ in (2) reduces to

$$T^P \widetilde{R}_{IJKL,M_1\cdots M_r P} = -(r+2)\widetilde{R}_{IJKL,M_1\cdots M_r}. \tag{6.4}$$

Proof. Differentiating the identity $T^K{}_{,I} = \delta^K{}_I$ shows that $T^K{}_{,IJ} = 0$. Commuting the derivatives shows that $T^L \widetilde{R}_{IJKL} = 0$. Now differentiating this relation successively and using again $T^L{}_{,M} = \delta^L{}_M$ proves (1).

The identity (2) can be proved as a consequence of (1) by commuting the contracted index to the left using the differentiated Ricci identity and then applying the second Bianchi identity and (1). It can alternately be derived directly as follows. Recall the general formula for the Lie derivative of a covariant tensor U in terms of a torsion-free connection:

$$(\mathcal{L}_X U)_{i\cdots j} = X^k U_{i\cdots j,k} + X^k{}_{,i} U_{k\cdots j} + \cdots + X^k{}_{,j} U_{i\cdots k}. \tag{6.5}$$

From the fact that $\mathcal{L}_T \widetilde{g} = 2\widetilde{g}$, it follows that $\mathcal{L}_T \widetilde{R}^{(r)} = 2\widetilde{R}^{(r)}$ for all r. Using this and $T^I{}_{,J} = \delta^I{}_J$ in (6.5), one concludes (6.4). Replacing r by s in (6.4) and differentiating $r - s$ more times using again $T^I{}_{,J} = \delta^I{}_J$ gives (2). □

Recall that $T^I = t\delta^I{}_0$. So a consequence of Proposition 6.1 is that t times a component of a tensor $\widetilde{R}^{(r+1)}$ with a 0 index can be expressed as a sum of components of $\widetilde{R}^{(r)}$ with the 0 index removed and with the remaining indices permuted. This implies a corresponding statement relating tensors $\widetilde{R}^{(r+1)}_{\mathcal{S}_0,\mathcal{S}_M,\mathcal{S}_\infty}$ and $\widetilde{R}^{(r)}_{\mathcal{S}'_0,\mathcal{S}'_M,\mathcal{S}'_\infty}$ on M.

When n is odd, for any choice of r and $\mathcal{S}_0$, $\mathcal{S}_M$, $\mathcal{S}_\infty$, the tensor $\widetilde{R}^{(r)}_{\mathcal{S}_0,\mathcal{S}_M,\mathcal{S}_\infty}$ depends only on g and each such tensor is a natural tensor invariant of g. However, as we have seen in the examples above, when n is even some of these tensors may depend on the specific chosen ambient metric and not just on g. The next result gives precise conditions for this not to happen.

Proposition 6.2. *Denote by s_0, s_M, s_∞ the cardinalities of the sets $\mathcal{S}_0$, $\mathcal{S}_M$, $\mathcal{S}_\infty$. Suppose n is even and $s_M + 2s_\infty \leq n+1$. Then the tensor $\widetilde{R}^{(r)}_{\mathcal{S}_0,\mathcal{S}_M,\mathcal{S}_\infty}$ is independent of the specific ambient metric which has been chosen and is a natural tensor invariant of g.*

Proof. Consider a component $\widetilde{R}_{IJKL,M_1\cdots M_r}$ for a straight ambient metric $\widetilde{g}$ in normal form relative to g. Let s_0, s_M, s_∞ denote the number of 0's, indices corresponding to M, and ∞'s, resp., in the list $IJKLM_1\cdots M_r$. We will prove by induction on r the following statement: all components of $\widetilde{R}^{(r)}$ satisfy that $\widetilde{R}_{IJKL,M_1\cdots M_r}$ mod $O(\rho^{(n+2-s_M-2s_\infty)/2})$ is independent of the $O(\rho^{n/2})$ ambiguity of the component $g_{ij}(x,\rho)$ of $\widetilde{g}$. Proposition 6.2 follows from this upon restricting to $\rho = 0$ and using the fact that the Taylor coefficients of $g_{ij}(x,\rho)$ of order $< n/2$ are natural tensor invariants of g.

We observe first that it suffices to assume that the power of ρ in our inductive statement satisfies $(n+2-s_M-2s_\infty)/2 \leq n/2-1$. Otherwise $s_M + 2s_\infty \leq 3$, which implies that at most 3 of the indices $IJKLM_1\cdots M_r$ are not equal to 0. Such a component $\widetilde{R}_{IJKL,M_1\cdots M_r}$ vanishes identically by Proposition 6.1.

Now proceed with the induction. The case $r=0$ follows easily from (6.1) and $\widetilde{R}_{IJK0} = 0$. For the inductive step, consider $\widetilde{R}_{IJKL,M_1\cdots M_rP}$. If $P=0$, the fact that the ambiguity of this component is no worse than that of $\widetilde{R}_{IJKL,M_1\cdots M_r}$ follows immediately from (6.4). For $P \neq 0$, write

$$\widetilde{R}_{IJKL,M_1\cdots M_rP} = \partial_P \widetilde{R}_{IJKL,M_1\cdots M_r} - \widetilde{\Gamma}^A_{IP}\widetilde{R}_{AJKL,M_1\cdots M_r} \\ - \ldots - \widetilde{\Gamma}^A_{M_rP}\widetilde{R}_{IJKL,M_1\cdots A}.$$

If $P \neq \infty$, then the ambiguity in the first term on the right-hand side vanishes to the same order as that in $\widetilde{R}_{IJKL,M_1\cdots M_r}$, while if $P=\infty$, then it vanishes to one order less. Thus the required vanishing for the first term follows from the inductive hypothesis. Consider next the second term. The Christoffel symbols are given by (3.16). They have their own ambiguity of at most $O(\rho^{(n/2-1)})$ owing to the $O(\rho^{n/2})$ ambiguity in $g_{ij}(x,\rho)$. But since by the observation above we can assume that $(n+2-s_M-2s_\infty)/2 \leq n/2-1$, we can neglect the ambiguity in the Christoffel symbols. Consider the ambiguity in the second term arising from the ambiguity in $\widetilde{R}_{AJKL,M_1\cdots M_r}$. If $P=\infty$, then we need to show that the order of vanishing of this ambiguity is at most 1 less than that of $\widetilde{R}_{IJKL,M_1\cdots M_r}$. But this is clear from the inductive hypothesis without even considering the Christoffel symbol since these two components of $\widetilde{R}^{(r)}$ have at most one different index and the change in ambiguity from changing one index is at most 1. If $P=p$ is between 1 and n, we need to show that the order of vanishing of the ambiguity in $\widetilde{\Gamma}^A_{Ip}\widetilde{R}_{AJKL,M_1\cdots M_r}$ is at most $\frac{1}{2}$ less than that of $\widetilde{R}_{IJKL,M_1\cdots M_r}$. This is clear by the same reasoning as in the case $P=\infty$ unless $I=0$ and $A=\infty$. But (3.16) shows that $\widetilde{\Gamma}^\infty_{0p}=0$, so that the inductive statement holds in this case also. The same reasoning as for the second term applies to the remaining terms on the right-hand side above. □

The weighting which appears in Proposition 6.2 suggests the following definition.

Definition 6.3. We define the *strength* of lists of indices in $\mathbb{R}^{n+2}$ as follows. Set $\|0\|=0$, $\|i\|=1$ for $1\le i\le n$, and $\|\infty\|=2$. For a list, write $\|I\dots J\|=\|I\|+\cdots+\|J\|$.

Proposition 6.2 thus asserts that for n even, the component $\widetilde{R}_{IJKL,M_1\cdots M_r}$ is well-defined at $\rho=0$ independent of the choice of ambient metric so long as $\|IJKLM_1\cdots M_r\|\le n+1$.

Next we consider the trace-free condition imposed by the asymptotic Ricci-flatness of ambient metrics.

Proposition 6.4. *If n is odd, the covariant derivatives of curvature of a straight ambient metric in normal form relative to g satisfy at $\rho=0$*

$$\widetilde{g}^{IK}\widetilde{R}_{IJKL,M_1\cdots M_r}=0. \tag{6.6}$$

If n is even, the same result holds assuming that $\|JLM_1\cdots M_r\|\le n-1$.

Proof. The result is clear for n odd since the Ricci curvature of $\widetilde{g}$ vanishes to infinite order. For n even we prove by induction on r the statement that

$$\widetilde{g}^{IK}\widetilde{R}_{IJKL,M_1\cdots M_r}=O(\rho^{(n-\|JLM_1\cdots M_r\|)/2})$$

for all components $JLM_1\cdots M_r$. The desired result then follows upon setting $\rho=0$.

The case $r=0$ of the induction is a consequence of the fact that $\widetilde{R}_{JL}=O^+_{JL}(\rho^{n/2-1})$. For the induction, write

$$\begin{aligned}\widetilde{g}^{IK}\widetilde{R}_{IJKL,M_1\cdots M_r}&=(\widetilde{g}^{IK}\widetilde{R}_{IJKL,M_1\cdots M_{r-1}})_{,M_r}\\&=\partial_{M_r}(\widetilde{g}^{IK}\widetilde{R}_{IJKL,M_1\cdots M_{r-1}})-\widetilde{\Gamma}^A_{JM_r}\widetilde{g}^{IK}\widetilde{R}_{IAKL,M_1\cdots M_{r-1}}\\&\quad-\dots-\widetilde{\Gamma}^A_{M_{r-1}M_r}\widetilde{g}^{IK}\widetilde{R}_{IJKL,M_1\cdots A}.\end{aligned}$$

The bound on the first term on the right-hand side follows from the induction hypothesis and the effect of differentiation on the order of vanishing. The bound on the remaining terms is easily seen to be a consequence of the induction hypothesis and the fact that $\widetilde{\Gamma}^I_{JK} = 0$ for $\|I\| > \|J\| + \|K\|$, which follows from (3.16). □

When the trace in (6.6) is written out in terms of components, the resulting identity can be interpreted as expressing a trace with respect to g of a conformal curvature tensor $\widetilde{R}^{(r)}_{\mathcal{S}_0,\mathcal{S}_M,\mathcal{S}_\infty}$ in terms of other conformal curvature tensors.

Suppose now we choose a conformally related metric $\widehat{g} = e^{2\Upsilon}g$. By Proposition 2.8, an ambient metric in normal form for g can be put into normal form for $\widehat{g}$ by a unique homogeneous diffeomorphism which restricts to the identity on $\mathcal{G}$. By calculating on $\mathcal{G}$ the Jacobian of this diffeomorphism and using the fact that the ambient curvature tensors are tensors, we will be able to compute the transformation laws of the conformal curvature tensors under conformal change. We denote the coordinates relative to $\widehat{g}$ by $(\widehat{t}, \widehat{x}, \widehat{\rho})$ and the conformal curvature tensors for $\widehat{g}$ by $\widehat{\widetilde{R}}^{(r)}_{\mathcal{S}_0,\mathcal{S}_M,\mathcal{S}_\infty} = \widehat{\widetilde{R}}_{IJKL,M_1\cdots M_r}|_{\widehat{\rho}=0,\,\widehat{t}=1}$.

Proposition 6.5. *Let g and $\widehat{g} = e^{2\Upsilon}g$ be conformally related metrics on M. Let $IJKLM_1\cdots M_r$ be a list of indices, s_0 of which are 0, s_M of which correspond to M, and s_∞ of which are ∞. If n is even, assume that $s_M + 2s_\infty \le n+1$. Then the conformal curvature tensors satisfy the conformal transformation law*

$$\widehat{\widetilde{R}}_{IJKL,M_1\cdots M_r}|_{\widehat{\rho}=0,\,\widehat{t}=1} = e^{2(1-s_\infty)\Upsilon}\widetilde{R}_{ABCD,F_1\cdots F_r}|_{\rho=0,\,t=1}p^A{}_I\cdots p^{F_r}{}_{M_r}, \tag{6.7}$$

where $p^A{}_I$ is the matrix

$$p^A{}_I = \begin{pmatrix} 1 & \Upsilon_i & -\frac{1}{2}\Upsilon_k\Upsilon^k \\ 0 & \delta^a{}_i & -\Upsilon^a \\ 0 & 0 & 1 \end{pmatrix}. \tag{6.8}$$

We make several observations and explanations before giving the proof. For each division of $IJKLM_1\cdots M_r$ into subsets $\mathcal{S}_0$, $\mathcal{S}_M$, and $\mathcal{S}_\infty$, the identity (6.7) is a relation amongst tensors on M. If Υ is constant, then $p^A{}_I = \delta^A{}_I$ and (6.7) just tells how $\widetilde{R}^{(r)}_{\mathcal{S}_0,\mathcal{S}_M,\mathcal{S}_\infty}$ scales. Because of the upper-triangular form of the matrix $p^A{}_I$, in the general case the other terms on the right-hand side all involve "earlier" conformal curvature tensors in the sense that each 'i' can be replaced only by 0 and each ∞ only by an 'i' or a 0. The conformal transformation law of a conformal curvature tensor involves only other conformal curvature tensors and first derivatives of Υ.

Consider the case $r = 0$. If we take s_0, $s_\infty = 0$, then $\widetilde{\widehat{R}}_{ijkl}|_{\widehat{\rho}=0,\,\widehat{t}=1} = \widehat{W}_{ijkl}$ is the Weyl tensor of $\widehat{g}$, and since $\widetilde{R}_{IJK0} = 0$, it follows that (6.7) reproduces the conformal invariance $\widehat{W}_{ijkl} = e^{2\Upsilon} W_{ijkl}$ of the Weyl tensor. Taking $I = \infty$ reproduces the transformation law $\widehat{C}_{jkl} = C_{jkl} - \Upsilon^i W_{ijkl}$ of the Cotton tensor, and taking I, $L = \infty$ gives for $n \neq 4$ the transformation law

$$\widehat{B}_{jk} = e^{-2\Upsilon}\left(B_{jk} + (n-4)\left[\Upsilon^l(C_{jkl} + C_{kjl}) - \Upsilon^i\Upsilon^l W_{ijkl}\right]\right)$$

of the Bach tensor. For $n = 4$ we have already noted the conformal invariance of the Bach tensor in its appearance as the obstruction tensor.

The conformal transformation law (6.7) is one of the most important features of the conformal curvature tensors and makes evident the importance of their interpretation as components of tensors on the ambient space. Since the conformal curvature tensors clearly vanish if g is flat, one consequence of (6.7) is that they also vanish if g is locally conformally flat. This is not a priori obvious since they are not individually conformally invariant. We will remove the restriction $s_M + 2s_\infty \leq n+1$ from this observation in even dimensions in Chapter 7. Collections of tensors on M which transform conformally according to rules of the form (6.7) define sections of weighted tensor powers of the cotractor bundle associated to the conformal structure. See [BEGo], [Go1] for development of this point of view, and [ČG], [BrG], [AL] for discussion of the relation between tractors and the ambient construction. Note that for each $x \in M$, the matrix $p^A{}_I(x)$ is in the orthogonal group of the quadratic form

$$\begin{pmatrix} 0 & 0 & 1 \\ 0 & g_{ij}(x) & 0 \\ 1 & 0 & 0 \end{pmatrix}.$$

Proof of Proposition 6.5. Let $\widetilde{g}$ be a straight ambient metric in normal form relative to g. The metric $\widetilde{g}$ is defined on a neighborhood of $\mathcal{G} \times \{0\}$ in $\mathcal{G} \times \mathbb{R}$. If we use the trivialization induced by g to identify $\mathcal{G}$ with $\mathbb{R}_+ \times M$, then $\widetilde{g}$ takes the form (3.14). According to Proposition 2.8, we can put $\widetilde{g}$ in normal form relative to $\widehat{g}$ by a unique homogeneous diffeomorphism ϕ defined on a neighborhood of $\mathcal{G} \times \{0\}$ which restricts to the identity on $\mathcal{G}$. If on the $\widehat{g}$ side we identify $\mathcal{G}$ with $\mathbb{R}_+ \times M$ using the trivialization induced by $\widehat{g}$, then $\phi^*\widetilde{g}$ will take the form (3.14) in coordinates $(\widehat{t}, \widehat{x}, \widehat{\rho})$. The trivializations are related by (2.1). Therefore we deduce the existence of a homogeneous diffeomorphism $\psi(\widehat{t}, \widehat{x}, \widehat{\rho}) = (t, x, \rho)$ on a homogeneous neighborhood of $\mathbb{R}_+ \times M \times \{0\}$ into $\mathbb{R}_+ \times M \times \mathbb{R}$ with the properties that $\widetilde{\widehat{g}} := \psi^*\widetilde{g}$ takes the form (3.14) in $(\widehat{t}, \widehat{x}, \widehat{\rho})$ and

$$\psi(\widehat{t}, \widehat{x}, 0) = (\widehat{t}e^{\Upsilon}, \widehat{x}, 0).$$

Now differentiation of this relation determines the derivatives of the components of ψ at $\widehat{\rho} = 0$ in the $\widehat{t}$ and $\widehat{x}$ directions. The derivatives in the $\widehat{\rho}$

direction are determined from these and from the requirement that $\psi^*\widetilde{g} = 2\widehat{t}d\widehat{t}d\widehat{\rho} + \widehat{t}^2\widehat{g}_{ij}d\widehat{x}^i d\widehat{x}^j$ at $\widehat{\rho} = 0$. This gives

$$(\psi')^A{}_I|_{\widehat{\rho}=0} = \begin{pmatrix} e^{\Upsilon} & \widehat{t}e^{\Upsilon}\Upsilon_i & -\frac{1}{2}\widehat{t}e^{-\Upsilon}\Upsilon_k\Upsilon^k \\ 0 & \delta^a{}_i & -e^{-2\Upsilon}\Upsilon^a \\ 0 & 0 & e^{-2\Upsilon} \end{pmatrix}.$$

Observe that this may be factored as

$$\psi'|_{\widehat{\rho}=0} = d_1 p d_2, \tag{6.9}$$

where p is given by (6.8) and

$$d_1 = \begin{pmatrix} \widehat{t}e^{\Upsilon} & 0 & 0 \\ 0 & Id & 0 \\ 0 & 0 & 1 \end{pmatrix} \qquad d_2 = \begin{pmatrix} \widehat{t}^{-1} & 0 & 0 \\ 0 & Id & 0 \\ 0 & 0 & e^{-2\Upsilon} \end{pmatrix}. \tag{6.10}$$

Each curvature component $\widetilde{R}_{IJKL,M_1\cdots M_r}$ is homogeneous with respect to the dilations. Since $\delta_s^*\widetilde{g} = s^2\widetilde{g}$, it follows that for the full tensors we have $\delta_s^*\widetilde{R}^{(r)} = s^2\widetilde{R}^{(r)}$. Since ∂_t is homogeneous of degree -1 and ∂_ρ and ∂_{x^i} are of degree 0, we deduce that the component $\widetilde{R}_{IJKL,M_1\cdots M_r}$ is homogeneous of degree $2 - s_0$. Therefore

$$\widehat{\widetilde{R}}_{IJKL,M_1\cdots M_r}|_{\widehat{\rho}=0,\,\widehat{t}=1} = e^{(2-s_0)\Upsilon}\widehat{\widetilde{R}}_{IJKL,M_1\cdots M_r}|_{\widehat{\rho}=0,\,\widehat{t}=e^{-\Upsilon}}. \tag{6.11}$$

Since the covariant derivatives of curvature are tensorial, we have

$$\widehat{\widetilde{R}}_{IJKL,M_1\cdots M_r} = \widetilde{R}_{ABCD,F_1\cdots F_r} \circ \psi\ (\psi')^A{}_I \cdots (\psi')^{F_r}{}_{M_r}.$$

Evaluate both sides at $\widehat{\rho} = 0$, $\widehat{t} = e^{-\Upsilon}$ and use (6.9) to obtain

$$\widehat{\widetilde{R}}_{IJKL,M_1\cdots M_r}|_{\widehat{\rho}=0,\widehat{t}=e^{-\Upsilon}} = e^{(s_0-2s_\infty)\Upsilon}\widetilde{R}_{ABCD,F_1\cdots F_r}|_{\rho=0,\,t=1}p^A{}_I \cdots p^{F_r}{}_{M_r}.$$

When combined with (6.11), this gives (6.7).

This argument is valid for all components of ambient curvature whether n is even or odd. However, when n is even, the components with $s_M + 2s_\infty > n+1$ generally depend on the ambient metric which has been chosen. To ensure that the component depends only on g as formulated in Proposition 6.5, we take $s_M + 2s_\infty \le n+1$ and use Proposition 6.2. □

If n is even and $\|IJKLM_1\cdots M_r\| \ge n+2$, then the restriction of $\widetilde{R}_{IJKL,M_1\cdots M_r}$ to $\rho = 0$ in general does depend on the choice of ambient metric. The component $\widetilde{R}_{\infty ij\infty,\underbrace{\infty\ldots\infty}_{n/2-2}}$ is an important example. The next proposition, which we will use in Chapter 7, makes explicit this dependence.

Proposition 6.6. *Let $n \ge 4$ be even. There is a natural trace-free symmetric 2-tensor K_{ij} depending on a pseudo-Riemannian metric g, with the following* *properties. Let $\widetilde{g}_{IJ}$ be an ambient metric in normal form relative to g, and write $\widetilde{g}_{ij} = t^2 g_{ij}(x,\rho)$. Then K_{ij} can be expressed algebraically in terms of the tensors $g^{ij}|_{\rho=0}$ and $\partial_\rho^m g_{ij}|_{\rho=0}$, $0 \le m \le n/2-1$, and one has*

$$2\widetilde{R}_{\infty ij\infty,\underbrace{\infty\ldots\infty}_{n/2-2}}|_{\rho=0,\,t=1} = \operatorname{tf}\left(\partial_\rho^{n/2} g_{ij}|_{\rho=0}\right) + K_{ij}. \tag{6.12}$$

Proof. First suppose that $\widetilde{g}$ is straight. An easy induction beginning with the last line of (6.1) and using (3.16) shows that for $m \ge 0$,

$$2t^{-2}\widetilde{R}_{\infty ij\infty,\underbrace{\infty\ldots\infty}_{m}} = \partial_\rho^{m+2} g_{ij} + \mathcal{P}_{ij}^{(m)}\left(g^{kl}, g_{kl}, \partial_\rho g_{kl}, \ldots, \partial_\rho^{m+1} g_{kl}\right), \tag{6.13}$$

where $\mathcal{P}_{ij}^{(m)}$ is a tensor depending polynomially on the indicated arguments. Define

$$K_{ij} = \operatorname{tf}\left(\mathcal{P}_{ij}^{(n/2-2)}|_{\rho=0}\right).$$

Using (3.2), we have

$$t^{-2} g^{ij}\widetilde{R}_{\infty ij\infty,\underbrace{\infty\ldots\infty}_{m}} = \widetilde{g}^{IJ}\widetilde{R}_{\infty IJ\infty,\underbrace{\infty\ldots\infty}_{m}} = -\widetilde{R}_{\infty\infty\infty,\underbrace{\infty\ldots\infty}_{m}}.$$

Since $\widetilde{R}_{IJ} = O(\rho^{n/2-1})$, it follows that $\widetilde{R}_{\infty ij\infty,\underbrace{\infty\ldots\infty}_{m}}|_{\rho=0,t=1}$ is trace-free for $m \le n/2-2$. So restricting (6.13) with $m = n/2-2$ to $\rho = 0$, $t = 1$ and taking the trace-free part gives the desired conclusion.

It is not hard to check using (3.2) and (3.3) that the last line of (6.1) and (6.13) are also valid even if $\widetilde{g}$ is not assumed to be straight, so that the same proof is valid in the general case. The $\widetilde{g}_{00}$ and $\widetilde{g}_{0i}$ components simply do not enter into any of the expressions which occur. □

Evaluating the last line of (6.1) using (3.6) shows that $K_{ij} = -2\operatorname{tf}\left(P_i{}^k P_{jk}\right)$ for $n = 4$.

The discussion in the rest of this chapter concerns the dimension dependence of certain 2-tensors. Consider partial contractions C of the form

$$\operatorname{pcontr}\left(\nabla^{r_1}R\otimes\cdots\otimes\nabla^{r_L}R\right) \quad \text{or} \quad \operatorname{pcontr}\left(\nabla^{r_1}R\otimes\cdots\otimes\nabla^{r_L}R\otimes g\right)$$

of the covariant derivatives of the curvature tensor of a metric g and possibly also g itself, in which 2 indices remain uncontracted. For partial contractions of the second type, we require that the uncontracted indices are the indices on g. Either type of partial contraction may be viewed formally as a choice of non-negative integers $r_1, \cdots, r_L$, a pairing of the contracted indices, and an ordering of the uncontracted indices. Consider also formal linear combinations $\xi = \sum_{i=1}^M f_i(d) C_i$ of such formal partial contractions whose coefficients $f_i(d)$ are rational functions of a single variable d. Any such formal linear combination ξ defines a natural 2-tensor $\mathrm{Eval}_{d=n}\,\xi$ in any dimension $n \geq 3$ which is a regular point for all the coefficients f_i by evaluation on metrics in dimension n. One can also define the residue of ξ at any dimension $n \geq 3$ to be the natural 2-tensor in dimension n given by

$$\mathrm{Res}_{d=n}\,\xi = \sum_{i=1}^{M} (\mathrm{Res}_{d=n}\, f_i)\, \mathrm{Eval}_{d=n}\, C_i.$$

By considering the product of a metric in a given dimension with flat metrics, it can be seen that if ξ satisfies that for some n_0, $\mathrm{Eval}_{d=n}\,\xi = 0$ for all $n \geq n_0$, then $\mathrm{Res}_{d=n}\,\xi = 0$ for all n. Thus the residue at any dimension is independent of the way a given dimension-dependent family of natural tensors is written formally.

The construction of the ambient metric in Chapter 3 shows that for each $m \geq 1$, there is a formal linear combination ξ_m as above such that if n is odd or if n is even and $m < n/2$, then for a straight ambient metric in dimension n in normal form relative to g, one has $\partial_\rho^m g_{ij}|_{\rho=0} = \mathrm{Eval}_{d=n}\,\xi_m$. For $m \geq 0$, the curvature component $\widetilde{R}_{\infty ij\infty,\underbrace{\infty\ldots\infty}_{m}}|_{\rho=0,t=1}$ can be so written as well for n odd or for n even and $m < n/2-2$. Fix an even integer $n \geq 4$ and consider $\mathrm{Res}_{d=n}\,\widetilde{R}_{\infty ij\infty,\underbrace{\infty\ldots\infty}_{n/2-2}}|_{\rho=0,t=1}$. As noted above, the residue is independent of the choice of formal expression, so this notation is justified. For $n = 4$ we have $\widetilde{R}_{\infty ij\infty}|_{\rho=0,t=1} = -(d-4)^{-1}B_{ij}$ from (6.2). Thus $\widetilde{R}_{\infty ij\infty}|_{\rho=0,t=1}$ has a simple pole at $d = 4$ with residue the negative of the Bach tensor. The following proposition generalizes this fact to higher n.

Proposition 6.7. *Let $n \geq 4$ be an even integer. $\widetilde{R}_{\infty ij\infty,\underbrace{\infty\ldots\infty}_{n/2-2}}|_{\rho=0,t=1}$ has a simple pole at $d = n$ with residue given by*

$$\mathrm{Res}_{d=n}\,\widetilde{R}_{\infty ij\infty,\underbrace{\infty\ldots\infty}_{n/2-2}}|_{\rho=0,t=1} = (-1)^{n/2-1}\left[2^{n/2-2}(n/2-2)!\right]^{-1}\mathcal{O}_{ij}.$$

Proof. Write $\widetilde{g}_{ij} = t^2 g_{ij}(x,\rho)$. The derivatives $\partial_\rho^m g_{ij}|_{\rho=0}$ are all regular in d at $d = n$ for $0 \leq m \leq n/2-1$, so the argument of Proposition 6.6 shows that

$$\mathrm{Res}_{d=n}\,\widetilde{R}_{\infty ij\infty,\underbrace{\infty\ldots\infty}_{n/2-2}}|_{\rho=0,t=1} = \tfrac{1}{2}\,\mathrm{Res}_{d=n}\,\mathrm{tf}\left(\partial_\rho^{n/2} g_{ij}|_{\rho=0}\right). \tag{6.14}$$

For $d > n$, $\mathrm{tf}\left(\partial_\rho^{n/2} g_{ij}|_{\rho=0}\right)$ is determined by replacing n by d in the first line of (3.17), setting the right-hand side equal to 0, applying $\partial_\rho^{n/2-1}|_{\rho=0}$, and taking the trace-free part. At $\rho = 0$ one has

$$\partial_\rho^{n/2-1}\left[\rho g''_{ij} - (d/2-1)g'_{ij}\right] = \tfrac{1}{2}(n-d)\partial_\rho^{n/2} g_{ij}.$$

For the remaining terms in the right-hand side of the first line of (3.17), set

$$T_{ij} = \mathrm{tf}\left[\partial_\rho^{n/2-1}\left(-\rho g^{kl}g'_{ik}g'_{jl} + \tfrac{1}{2}\rho g^{kl}g'_{kl}g'_{ij} - \tfrac{1}{2}g^{kl}g'_{kl}g_{ij} + R_{ij}\right)|_{\rho=0}\right].$$

Then T_{ij} is expressible in terms of the $\partial_\rho^m g_{ij}|_{\rho=0}$ for $m \le n/2-1$, so is regular at $d = n$. Differentiation of (3.17) as indicated above thus gives $\frac{1}{2}(n-d)\,\mathrm{tf}\left(\partial_\rho^{n/2}g_{ij}|_{\rho=0}\right) + T_{ij} = 0$, so

$$\tfrac{1}{2}\,\mathrm{Res}_{d=n}\,\mathrm{tf}\left(\partial_\rho^{n/2}g_{ij}|_{\rho=0}\right) = T_{ij}|_{d=n}. \tag{6.15}$$

On the other hand, the obstruction tensor $\mathcal{O}_{ij}$ in dimension n is obtained by setting $\widetilde{R}_{ij} = c_n^{-1}(2\rho)^{n/2-1}\mathcal{O}_{ij} \mod O(\rho^{n/2})$ in the first line of (3.17) (without changing n to d), applying $\partial_\rho^{n/2-1}|_{\rho=0}$, and taking the trace-free part. This gives

$$c_n^{-1}2^{n/2-1}(n/2-1)!\,\mathcal{O}_{ij} = T_{ij}|_{d=n}.$$

Combining with (6.15) and (6.14) gives the result. □

Thus $(d-n)\widetilde{R}_{\infty ij\infty,\underbrace{\infty\ldots\infty}_{n/2-2}}|_{\rho=0,t=1}$ is a natural trace-free symmetric 2-tensor depending on d regular also at $d=n$ which equals a multiple of the obstruction tensor for $d=n$. Of course, there are many other such tensors; one can add to $\widetilde{R}_{\infty ij\infty,\underbrace{\infty\ldots\infty}_{n/2-2}}|_{\rho=0,t=1}$ any tensor regular at $d=n$. For example, one can replace $\widetilde{R}_{\infty ij\infty,\underbrace{\infty\ldots\infty}_{n/2-2}}|_{\rho=0,t=1}$ by $\frac{1}{2}\,\mathrm{tf}\left(\partial_\rho^{n/2}g_{ij}|_{\rho=0}\right)$; the relation (6.12) holds also as dimension-dependent natural tensors in dimensions d which are odd or which are even and $> n$, with K_{ij} regular also at $d=n$. A main advantage of the choice $\widetilde{R}_{\infty ij\infty,\underbrace{\infty\ldots\infty}_{n/2-2}}|_{\rho=0,t=1}$ is that its transformation law under conformal change is known and relatively simple, being given by (6.7) for d odd or $d>n$ even.

One can also consider the continuation of $\widetilde{R}_{\infty ij\infty,\underbrace{\infty\ldots\infty}_{n/2-2}}|_{\rho=0,t=1}$ to even values of $d<n$. In general, the pole is not simple and its order increases with $n-d$.

Chapter Seven

Conformally Flat and Conformally Einstein Spaces

If n is odd, an ambient metric in normal form is uniquely determined to infinite order by (M, g). Theorem 3.9 shows that within the family of all formal solutions, this one is distinguished by the vanishing of the $\rho^{n/2}$ coefficient, or equivalently by the condition that it be smooth. In the Poincaré realization, the corresponding condition is that the expansion have no r^n term, or equivalently that it be even. The fact that there is such a normalization giving rise to a unique diffeomorphism class of ambient metrics associated to the conformal manifold $(M, [g])$ is crucial for the applications to jet isomorphism and invariant theory in Chapters 8 and 9.

When n is even, the situation is different primarily because of the existence of the obstruction tensor, which gives rise to the log terms in the infinite order expansions determined in Theorem 3.10. But even for conformal classes with vanishing obstruction tensor, for which all formal solutions are smooth, the coefficient of $\rho^{n/2}$ is undetermined, so there is not a unique solution. One can impose the condition that $\widetilde{g}$ be straight; this removes some of the freedom, but there is no apparent analog of the odd-dimensional normalization.

In this chapter we discuss two families of special conformal classes for which there is a natural conformally invariant normalization in even dimensions giving rise to a unique diffeomorphism class of infinite order formal expansions for an ambient (or Poincaré) metric. These are the locally conformally flat conformal classes and the conformal classes containing an Einstein metric.

Consider first the case of a locally conformally flat manifold $(M, [g])$. This means that in a neighborhood of any point of M, there is a flat metric, say γ, in the conformal class. Now (6.1) implies that the metric $2\rho dt^2 + 2t dt d\rho + t^2\gamma$, with γ independent of ρ, is flat. If n is odd, this implies that every ambient metric is flat to infinite order, since the ambient metric is unique to infinite order up to diffeomorphism and the condition that $\widetilde{g}$ is flat is invariant under diffeomorphism. If $n \geq 4$ is even, the condition that $\widetilde{g}$ be flat to infinite order uniquely determines $\widetilde{g}$ to infinite order up to diffeomorphism. This follows from Theorem 3.10 and Proposition 6.6: the obstruction tensor vanishes and Theorem 3.10 shows that the full expansion is determined by $\operatorname{tf}\left(\partial_\rho^{n/2} g_{ij}|_{\rho=0}\right)$, which by Proposition 6.6 must equal $-K_{ij}$ if $\widetilde{g}$ is flat. Thus we have:

Proposition 7.1. *Let $n \geq 3$ and suppose that $(M, [g])$ is locally conformally flat. Then there exists an ambient metric $\tilde{g}$ for $(M, [g])$ which is flat to infinite order, and such a $\tilde{g}$ is unique to infinite order up to diffeomorphism.*

We remark that it follows from the fact that the metric $2\rho dt^2 + 2tdtd\rho + t^2\gamma$ is straight, that $\tilde{g}$ in Proposition 7.1 can be taken to be straight. In Proposition 7.2 and Theorem 7.4, we will extend Proposition 7.1 to establish the existence and uniqueness of a flat ambient metric not just to infinite order, but in a neighborhood of $\mathcal{G} \times \{0\}$.

One consequence of Proposition 7.1 is that the conformal curvature tensors can be invariantly defined for all orders of differentiation for locally conformally flat metrics in even dimensions. Because of conformal invariance, Proposition 6.5 continues to hold, so all the conformal curvature tensors are defined and vanish for locally conformally flat metrics.

When $n = 2$ the situation is quite different. (Recall that when $n = 2$, every metric is locally conformally flat.) In fact, when $n = 2$, every straight ambient metric is flat to infinite order. This follows by consideration of the Poincaré metric. If $\tilde{g}$ is a straight ambient metric, then $\mathrm{Ric}(\tilde{g})$ vanishes to infinite order, so Proposition 4.7 shows that the Poincaré metric g_+ defined in Proposition 4.6 satisfies that $\mathrm{Ric}(g_+) + ng_+$ vanishes to infinite order. Since the Ricci tensor determines the curvature tensor in dimension 3, it follows that $\mathrm{Rm}(g_+) + g_+ \owedge g_+$ vanishes to infinite order, so by Proposition 4.7 again, $\tilde{g}$ is flat to infinite order. There are many diffeomorphism-inequivalent straight ambient metrics for $(M, [g])$; they are parametrized in Theorem 3.7.

An explicit formula for flat ambient metrics in normal form relative to an arbitrary metric in the conformal class was given in [SS] (in the Poincaré realization). This is closely related to a result of Epstein [E1] describing the form of hyperbolic metrics near conformal infinity in terms of data on an interior hypersurface. In the following, we will denote by γ an arbitrary metric in the conformal class on M, and by $g = g_\rho$ a metric obtained by fixing ρ in the ij component of an ambient metric. The formula of [SS] is as follows.

Proposition 7.2. *Let γ_{ij} be a locally conformally flat metric on a manifold M of dimension n and let P_{ij} be a symmetric 2-tensor on M. Set*

$$(g_\rho)_{ij} = \gamma_{ij} + 2P_{ij}\rho + P_{ik}P^k{}_j\rho^2 \tag{7.1}$$

and define a metric $\tilde{g}$ on $\mathbb{R}_+ \times U$, where U is the subset of $M \times \mathbb{R}$ on which g_ρ is nondegenerate, by

$$\tilde{g} = 2tdtd\rho + 2\rho dt^2 + t^2 g_\rho.$$

Then $\tilde{g}$ is flat if:

- *$n \geq 3$ and P is the Schouten tensor of γ.*
- *$n = 2$ and P is any symmetric 2-tensor satisfying*

$$P_i{}^i = \tfrac{1}{2}R \qquad \text{and} \qquad P_{ij,}{}^j = \tfrac{1}{2}R_{,i}.$$

Here indices are raised and lowered using γ, indices preceded by a comma denote covariant derivatives with respect to the Levi-Civita connection of γ, and R denotes the scalar curvature of γ.

Note that the local conformal flatness of γ implies for $n \geq 3$ that the Schouten tensor of γ satisfies $P_{ij,k} = P_{(ij,k)}$. For $n = 2$, the hypotheses on P imply that $P_{i[j,k]}$ is trace-free. But in two dimensions, a trace-free tensor A_{ijk} satisfying $A_{ijk} = A_{i[jk]}$ and $A_{[ijk]} = 0$ must vanish identically (see [W]). So in all dimensions, a consequence of the hypotheses of Proposition 7.2 is that $P_{ij,k} = P_{(ij,k)}$.

As described in Propositions 4.6 and 4.7, under the change of variable $\rho = -\frac{1}{2}r^2$, the condition that $\widetilde{g}$ is flat is equivalent to the condition that g_+ is hyperbolic (i.e., has constant sectional curvature -1), where

$$g_+ = r^{-2}\left(dr^2 + g_{-\frac{1}{2}r^2}\right).$$

Thus one obtains explicit formulae for hyperbolic metrics. Consequences for the coefficients in the expansion of the volume form of g_+ in the case of a locally conformally flat infinity are given in [GJ].

We will indicate three proofs of Proposition 7.2. First we prove it by direct calculation.

Since there are two metrics involved, it is important to be careful about conventions for raising and lowering indices. We will use γ to raise and lower indices with the one exception that g^{ij} denotes the inverse of g_{ij}. Hereafter we write g for g_ρ; the ρ dependence is to be understood. We can write

$$g_{ij} = \gamma_{il}U^l{}_kU^k{}_j = U_{ik}U^k{}_j$$

where

$$U^i{}_j = \delta^i{}_j + \rho P^i{}_j. \tag{7.2}$$

Note that g is nondegenerate precisely where U is nonsingular. Let $V = U^{-1}$ so that $V^i{}_kU^k{}_j = \delta^i{}_j$ and observe that $U_{ij} = U_{ji}$, $V_{ij} = V_{ji}$. It is evident that

$$V^k{}_ig_{kj} = U_{ij} \tag{7.3}$$

and

$$g'_{ij} = 2P_{ik}U^k{}_j. \tag{7.4}$$

First we derive the relation between the Levi-Civita connections ${}^g\nabla$ and ${}^\gamma\nabla$ and the curvature tensors ${}^gR_{ijkl}$ and ${}^\gamma R_{ijkl}$ of g and γ.

Lemma 7.3. *Let g be given by (7.1), where γ is a metric, P_{ij} is a symmetric 2-tensor satisfying $P_{ij,k} = P_{(ij,k)}$ and $\rho \in \mathbb{R}$. If g is nondegenerate, then the Levi-Civita connections are related by*

$$^{g}\nabla_i \eta_j = {}^{\gamma}\nabla_i \eta_j - \rho V^k{}_l P^l{}_{i,j} \eta_k \tag{7.5}$$

and the curvature tensors by

$$^{g}R_{ijkl} = {}^{\gamma}R_{abkl} U^a{}_i U^b{}_j. \tag{7.6}$$

Proof. The right-hand side of (7.5) defines a torsion-free connection, so for (7.5) it suffices to show that g is parallel. Differentiating (7.1) gives

$$^{\gamma}\nabla_k g_{ij} = 2\rho P_{ij,k} + 2\rho^2 P^l{}_{(i} P_{j)l,k}. \tag{7.7}$$

On the other hand, using (7.3), (7.2) and the symmetry of $P_{ij,k}$ gives

$$V^m{}_l P^l{}_{k,(i} g_{j)m} = P^l{}_{k,(i} U_{j)l} = P_{ij,k} + \rho P^l{}_{k,(i} P_{j)l}.$$

Combining these shows that g is parallel.

Set $D^i{}_{jk} = \rho V^i{}_l P^l{}_{j,k}$. The difference of the curvature tensors of the connections is given in terms of the difference tensor by

$$^{g}R_{mjkl} g^{im} = {}^{\gamma}R^i{}_{jkl} + 2D^i{}_{j[l,k]} + 2D^c{}_{j[l} D^i{}_{k]c}.$$

Now

$$\begin{aligned} D^i{}_{jl,k} &= \rho \left(V^i{}_{a,k} P^a{}_{j,l} + V^i{}_a P^a{}_{j,lk} \right) \\ &= -\rho V^i{}_b U^b{}_{c,k} V^c{}_a P^a{}_{j,l} + \rho V^i{}_a P^a{}_{j,lk} \\ &= -\rho^2 V^i{}_b P^b{}_{c,k} V^c{}_a P^a{}_{j,l} + \rho V^i{}_a P^a{}_{j,lk} \\ &= -D^i{}_{ck} D^c{}_{jl} + \rho V^i{}_a P^a{}_{j,lk} \end{aligned}$$

so

$$\begin{aligned} ^{g}R_{mjkl} g^{im} &= {}^{\gamma}R^i{}_{jkl} + 2\rho V^{ia} P_{aj,[lk]} \\ &= {}^{\gamma}R^i{}_{jkl} + \rho V^{ia} \left({}^{\gamma}R^b{}_{alk} P_{bj} + {}^{\gamma}R^b{}_{jlk} P_{ab} \right). \end{aligned}$$

Using (7.3), we obtain

$$\begin{aligned} ^{g}R_{ijkl} &= {}^{\gamma}R^b{}_{jkl} g_{bi} + \rho U^a{}_i \left({}^{\gamma}R^b{}_{alk} P_{bj} + {}^{\gamma}R^b{}_{jlk} P_{ab} \right) \\ &= {}^{\gamma}R^b{}_{jkl} U_{ba} U^a{}_i + \rho U^a{}_i \left({}^{\gamma}R^b{}_{alk} P_{bj} + {}^{\gamma}R^b{}_{jlk} P_{ab} \right) \\ &= \left[{}^{\gamma}R^b{}_{jkl} \left(U_{ab} - \rho P_{ab} \right) + {}^{\gamma}R^b{}_{alk} (\rho P_{bj}) \right] U^a{}_i \\ &= \left[{}^{\gamma}R^b{}_{jkl} \gamma_{ab} + {}^{\gamma}R_{balk} (\rho P^b{}_j) \right] U^a{}_i \\ &= \left[{}^{\gamma}R_{ajkl} + {}^{\gamma}R_{abkl} (\rho P^b{}_j) \right] U^a{}_i \\ &= \left[{}^{\gamma}R_{abkl} \left(\delta^b{}_j + \rho P^b{}_j \right) \right] U^a{}_i \\ &= {}^{\gamma}R_{abkl} U^b{}_j U^a{}_i, \end{aligned}$$

which is (7.6). □

We remark that (7.6) can be viewed as a compatibility condition on a solution P_{ij} of the system $P_{i[j,k]} = 0$, which is overdetermined if $n \geq 3$.

Proof of Proposition 7.2. According to (6.1), $\widetilde{g}$ is flat if and only if

$$^gR_{ijkl} + \frac{1}{2}(g_{il}g'_{jk} + g_{jk}g'_{il} - g_{ik}g'_{jl} - g_{jl}g'_{ik}) + \frac{\rho}{2}(g'_{ik}g'_{jl} - g'_{il}g'_{jk}) = 0 \quad (7.8)$$

$$^g\nabla_k g'_{ij} - {}^g\nabla_j g'_{ik} = 0 \quad (7.9)$$

$$g''_{ij} - \frac{1}{2}g^{pq}g'_{ip}g'_{jq} = 0. \quad (7.10)$$

We will verify (7.8)–(7.10) by direct calculation.

Begin with (7.10). Differentiating (7.1) gives $g''_{ij} = 2P_{ik}P^k{}_j$. Now $g^{ij} = V^{il}V_l{}^j$, so using (7.4) gives $g^{pq}g'_{ip}g'_{jq} = 4P_{ik}P^k{}_j$. This proves (7.10).

Next consider (7.9). Differentiating (7.7) with respect to ρ gives

$$^\gamma\nabla_k g'_{ij} = 2P_{ij,k} + 4P^l{}_{(i}P_{j)l,k}\,\rho.$$

Using (7.4) gives

$$V^m{}_l P^l{}_{k,(i}g'_{j)m} = 2V^m{}_l P^l{}_{k,(i}P_{j)a}U^a{}_m = 2P^l{}_{k,(i}P_{j)l}.$$

Applying (7.5) and combining terms gives $^g\nabla_k g'_{ij} = 2P_{ij,k}$, so (7.9) holds.

To prove (7.8), note first that an easy calculation shows that

$$g_{ij} - \tfrac{1}{2}\rho g'_{ij} = U_{ij}.$$

Thus we obtain

$$\begin{aligned} g_{i[l}g'_{k]j} + g_{j[k}g'_{l]i} - \rho g'_{i[l}g'_{k]j} &= \left(g - \tfrac{1}{2}\rho g'\right)_{i[l}\, g'_{k]j} + \left(g - \tfrac{1}{2}\rho g'\right)_{j[k}\, g'_{l]i} \\ &= U_{i[l}g'_{k]j} + U_{j[k}g'_{l]i}. \end{aligned}$$

Substituting (7.4) gives

$$g_{i[l}g'_{k]j} + g_{j[k}g'_{l]i} - \rho g'_{i[l}g'_{k]j} = 2U_i{}^aU_j{}^b\left(\gamma_{a[l}P_{k]b} + \gamma_{b[k}P_{l]a}\right).$$

But the hypotheses of Proposition 7.2 imply that

$$^\gamma R_{abkl} = -2\left(\gamma_{a[l}P_{k]b} + \gamma_{b[k}P_{l]a}\right)$$

in all dimensions: this is automatically true for $n = 3$, is true for $n \geq 4$ because γ is locally conformally flat, and is true for $n = 2$ because $P_i{}^i = \frac{1}{2}R$ and the space of algebraic curvature tensors is one-dimensional. (7.8) follows upon combining with (7.6). □

We next sketch another proof of Proposition 7.2 suggested by the discussion in [SS] which avoids the calculations of Lemma 7.3. One first verifies (7.10) exactly as above, which is very simple. According to (6.1), this means that $\widetilde{R}_{\infty ij\infty} = 0$. Now consider the Bianchi identity $\widetilde{R}_{\infty i[jk,\infty]} = 0$. Write out the skew-symmetrization and then write the covariant derivatives as coordinate derivatives plus Christoffel symbol terms and use (3.16). Since $\widetilde{R}_{\infty ij\infty} = 0$, $\widetilde{R}_{IJK0} = 0$ and $\widetilde{\Gamma}^A_{\infty\infty} = 0$, the only components of curvature which appear are $\widetilde{R}_{\infty ijk}$. One concludes that

$$\partial_\rho \widetilde{R}_{\infty ijk} + A^{pqr}{}_{ijk}\widetilde{R}_{\infty pqr} = 0,$$

where $A^{pqr}{}_{ijk}$ is smooth. For each point of M, this is a linear system of ordinary differential equations in ρ, so $\widetilde{R}_{\infty ijk} = 0$ so long as this holds at $\rho = 0$, which follows easily from the hypotheses on P_{ij}. Finally one applies the same argument to the Bianchi identity $\widetilde{R}_{ij[kl,\infty]} = 0$ to deduce that $\widetilde{R}_{ijkl} = 0$.

A consequence of Proposition 7.2 is the fact that for $(M, [\gamma])$ locally conformally flat, an ambient metric can be chosen which is flat in a neighborhood of $\mathcal{G} \times \{0\}$ (as opposed to just to infinite order), or equivalently that a Poincaré metric can be chosen which is hyperbolic.

The arguments in [SS] and [E1] used to derive the form of a hyperbolic Poincaré metric also give uniqueness of the solution up to diffeomorphism in an open set, not just to infinite order. This may be of some interest in hyperbolic geometry, particularly the distinction between the cases $n = 2$ and $n > 2$. We formulate the Poincaré metric version rather than the ambient metric version of the result.

Theorem 7.4. *Let γ be a smooth metric on a manifold M, and let g_+ be a hyperbolic Poincaré metric defined on M_+°, where M_+ is a neighborhood of $M \times \{0\}$ in $M \times [0, \infty)$, with conformal infinity $(M, [\gamma])$. Then there is a neighborhood $\mathcal{U}$ of $M \times \{0\}$ in $M \times [0, \infty)$ and a diffeomorphism ϕ mapping $\mathcal{U}$ into M_+ which restricts to the identity on $M \times \{0\}$, so that*

$$\phi^* g_+ = r^{-2}\left(dr^2 + g_r\right), \tag{7.11}$$

where

$$(g_r)_{ij} = \gamma_{ij} - P_{ij}r^2 + \tfrac{1}{4}P_{ik}P^k{}_j r^4$$

and

- *If $n \geq 3$, P is the Schouten tensor of γ.*
- *If $n = 2$, P is some symmetric 2-tensor on M satisfying*

$$P_i{}^i = \tfrac{1}{2}R \qquad \text{and} \qquad P_{ij,}{}^j = \tfrac{1}{2}R_{,i}.$$

Proof. Take $\mathcal{U}$, ϕ as in Proposition 4.3, so that $\phi^* g_+$ is in normal form relative to γ. Since g_+ is hyperbolic, the ambient metric determined by $\phi^* g_+$

by the change of variable $\rho = -\frac{1}{2}r^2$ is flat and in normal form. Its curvature is given by (6.1). In particular, we have (7.10). Differentiating (7.10) with respect to ρ and substituting (7.10) for the second derivatives which occur gives $g'''_{ij} = 0$, where $' = \partial_\rho$. Thus g_{ij} is a quadratic polynomial in ρ.

We saw in Chapter 6 that for $n \geq 3$ the Weyl and Cotton tensors of γ can be recovered by restricting the curvature tensor of $\widetilde{g}$ to $\rho = 0$. Since $\widetilde{g}$ is flat, γ is locally conformally flat. The uniqueness parts of Proposition 7.1 and Theorem 3.7 now show that the ρ and ρ^2 coefficients of g_{ij} must be of the form given in Proposition 7.2. □

In particular, Theorem 7.4 implies that if $n \geq 3$, any two conformally compact hyperbolic metrics with the same conformal infinity are isometric in a neighborhood of the boundary by a diffeomorphism which restricts to the identity on the boundary. The same result holds for $n = 2$, except that one must require not only that the conformal infinities agree but that the r^2 coefficients P_{ij} agree as well. The proof of Theorem 7.4 is valid with weaker regularity hypotheses than our usual assumption that the compactification of g_+ is infinitely differentiable. There is an equivalent statement of Theorem 7.4 in terms of flat ambient metrics, whose formulation we leave to the reader.

Finally, we note that the argument of Theorem 7.4 can be used to give yet another proof of Proposition 7.2. If $n \geq 3$ and γ is locally conformally flat, Proposition 7.1 shows that there exists a straight ambient metric for $(M, [\gamma])$ which is flat to infinite order. Put this metric into normal form relative to γ. The argument of Theorem 7.4 shows that modulo terms vanishing to infinite order, the corresponding g_ρ must be a quadratic polynomial in ρ and its coefficients must be given by (7.1). If one takes g_ρ to equal this quadratic polynomial, then $\widetilde{g}$ is real analytic in ρ. Since its curvature tensor vanishes to infinite order at $\rho = 0$, it must vanish identically, so $\widetilde{g}$ is flat as desired. The same argument applies for $n = 2$ as well, using Theorem 3.7 and the discussion after Proposition 7.1 to conclude the existence of a straight ambient metric having the prescribed initial asymptotics which is flat to infinite order.

Consider next the case of conformal classes $[g]$ containing an Einstein metric. For this discussion we assume $n \geq 3$. We already noted in Chapter 3 that if g is Einstein, with $R_{ij} = 2\lambda(n-1)g_{ij}$ so that $P_{ij} = \lambda g_{ij}$, then

$$\widetilde{g} = 2\rho dt^2 + 2t dt d\rho + t^2 g_\rho, \qquad g_\rho = (1 + \lambda\rho)^2 g, \tag{7.12}$$

is straight, Ricci-flat, and in normal form relative to g. In the Poincaré realization, this is a reformulation of the familiar warped product construction of an Einstein metric in $n+1$ dimensions from an Einstein metric in n dimensions (see for example [Be]): for $\lambda \neq 0$, under the change of variable

$r = \sqrt{-2\rho} = \sqrt{\frac{2}{|\lambda|}}e^{-v}$, the associated Poincaré metric becomes

$$g_+ = r^{-2}\left(dr^2 + (1 - \tfrac{1}{2}\lambda r^2)^2 g\right) = dv^2 + 2|\lambda|\,\psi(v)\,g, \tag{7.13}$$

where

$$\psi(v) = \begin{cases} \sinh^2 v & \text{if } \lambda > 0 \\ \cosh^2 v & \text{if } \lambda < 0. \end{cases}$$

For $\lambda = 0$, the corresponding change of variable is $r = e^{-v}$, giving $g_+ = dv^2 + e^{2v}g$. See also [GrH1], [Leit], [Leis], [Ar]. The explicit representation (7.12) has been generalized to the case of certain products of Einstein metrics in [GoL]. See also the remarkable examples of Nurowski [N] of explicit ambient metrics.

When n is odd, then up to diffeomorphism (7.12) is the unique ambient metric to infinite order. But when n is even, there are others satisfying $\operatorname{Ric}(\widetilde{g}) = O(\rho^\infty)$, corresponding to other choices of tf $\left(\partial_\rho^{n/2} g_{ij}|_{\rho=0}\right)$ in Theorem 3.10. The following result shows that modulo diffeomorphism, the ambient metric (7.12) is uniquely determined to infinite order by the conformal class alone.

Proposition 7.5. *Let g be an Einstein metric and let $\widetilde{g}$ be the ambient metric defined by* (7.12). *Suppose that a conformal rescaling $\widehat{g} = e^{2\Upsilon}g$ is also Einstein with constant $\widehat{\lambda}$, and let $\widehat{\widetilde{g}}$ denote the ambient metric defined by taking $g_\rho = (1 + \widehat{\lambda}\rho)^2\widehat{g}$ in* (7.12). *Then $\widetilde{g}$ and $\widehat{\widetilde{g}}$ are diffeomorphic to infinite order.*

Of course, the Poincaré metrics (7.13) determined by g and $\widehat{g}$ are diffeomorphic to infinite order as well. As mentioned in the introduction, Proposition 7.5 was obtained by Haantjes-Schouten.

We begin the proof of Proposition 7.5 by characterizing the ambient metric (7.12) in terms of its curvature.

Proposition 7.6. *Let g be Einstein and let $\widetilde{g}$ be defined by* (7.12). *The covariant derivatives of curvature of $\widetilde{g}$ satisfy $\widetilde{R}_{IJKL,M\cdots N} = 0$ if at most 3 of the indices $IJKLM\cdots N$ are between 1 and n.*

Proof. First, substituting g_ρ into (6.1) gives $\widetilde{R}_{\infty jkl} = 0$ and $\widetilde{R}_{\infty ij\infty} = 0$. Now calculating the covariant derivatives inductively using the observations from (3.16) that $\widetilde{\Gamma}^A_{\infty\infty} = 0$ for all A and $\widetilde{\Gamma}^A_{i\infty} = 0$ unless $1 \le A \le n$, one finds that $\widetilde{R}_{\infty ij\infty,\infty\cdots\infty} = 0$ and $\widetilde{R}_{\infty ijk,\infty\cdots\infty} = 0$. Also, recalling $\widetilde{R}_{0JKL} = 0$ and calculating gives $\widetilde{R}_{\infty ij\infty,k} = 0$. Inductively differentiating this relation shows that $\widetilde{R}_{\infty ij\infty,k\infty\cdots\infty} = 0$.

Consider now $\widetilde{R}_{IJKL,M\cdots N}$ with at most 3 indices between 1 and n. Recall from Proposition 6.1 that a 0 index can be removed at the expense of permuting the remaining indices. Thus it can be assumed that none of the indices is 0. The symmetries of the curvature tensor then show that $\widetilde{R}_{IJKL,M\cdots N}$ vanishes unless at least two of $IJKL$ are between 1 and n. Thus at most one of $M\cdots N$ can be between 1 and n. We have shown above that all such components vanish, except for components $\widetilde{R}_{\infty ij\infty,\infty\cdots\infty k\infty\cdots\infty}$ with at least one ∞ to the right of the comma before the k. This can be shown to vanish by commuting the k to the left: by the differentiated Ricci identity, one has that

$$\widetilde{R}_{\infty ij\infty,\underbrace{\infty\cdots\infty}_{a}\infty k\underbrace{\infty\cdots\infty}_{b}} - \widetilde{R}_{\infty ij\infty,\underbrace{\infty\cdots\infty}_{a}k\infty\underbrace{\infty\cdots\infty}_{b}}$$

is a sum of terms of the form

$$\widetilde{R}^{P}{}_{M\infty k,\infty\cdots\infty}\widetilde{R}_{\dots,\dots\dots}$$

where M is one of i, j, ∞ and P contracts against one of the indices of the second factor. But it follows from what we have already shown that $\widetilde{R}_{QM\infty k,\infty\cdots\infty} = 0$ for all choices of M and Q. $\square$

In particular, for n even one has

$$\widetilde{R}_{\infty ij\infty,\underbrace{\infty\cdots\infty}_{n/2-2}} = 0 \quad \text{at } \rho = 0. \tag{7.14}$$

Recall from Proposition 6.6 that this condition determines $\operatorname{tf}\left(\partial_\rho^{n/2} g_{ij}|_{\rho=0}\right)$ (which clearly vanishes in this case, so that K_{ij} in Proposition 6.6 vanishes for an Einstein metric), and by Theorem 3.10 this determines the solution to infinite order. Thus we have:

Proposition 7.7. *Let $n \geq 4$ be even. If g is an Einstein metric, then $\widetilde{g}$ given by* (7.12) *is to infinite order the unique ambient metric in normal form relative to g satisfying* $\operatorname{Ric}(\widetilde{g}) = O(\rho^\infty)$ *and* (7.14).

The proof of Proposition 7.5 uses relations between the curvature of an Einstein metric g and the conformal factor relating g to another Einstein metric. Let g be Einstein and let $\widetilde{g}$ be defined by (7.12). Recall from Chapter 6 that if $r \geq 0$ and if we divide the indices $IJKLM_1\cdots M_r$ into three disjoint subsets $\mathcal{S}_0$, $\mathcal{S}_M$, $\mathcal{S}_\infty$ of cardinalities s_0, s_M, s_∞, resp., then we can define a covariant tensor $R_{\mathcal{S}_0,\mathcal{S}_M,\mathcal{S}_\infty}$ on M of rank s_M as follows. In the derivative of ambient curvature $\widetilde{R}_{IJKL,M_1\cdots M_r}|_{\rho=0,t=1}$, set the indices in $\mathcal{S}_0$ to be 0's, the indices in $\mathcal{S}_\infty$ to be ∞'s, and let those in $\mathcal{S}_M$ correspond to M in the identification $\mathcal{G}\times\mathbb{R} \cong \mathbb{R}_+ \times M \times \mathbb{R}$ induced by g.

Proposition 7.8. *Let g and $\widehat{g} = e^{2\Upsilon} g$ be conformally related Einstein metrics. Divide the indices $IJKLM_1 \cdots M_r$ into three disjoint subsets $\mathcal{S}_0$, $\mathcal{S}_M$, $\mathcal{S}_\infty$ and construct the tensor $\widetilde{R}_{\mathcal{S}_0,\mathcal{S}_M,\mathcal{S}_\infty}$ on M as described above. Now divide $\mathcal{S}_M$ into two disjoint subsets: $\mathcal{S}_M = \mathcal{T}_1 \cup \mathcal{T}_2$, and set* $\operatorname{card}(\mathcal{T}_i) = t_i$, *$i = 1, 2$. Construct a covariant tensor Y on M of rank t_2 by contracting the vector field* $\operatorname{grad}\Upsilon$ *into each of the indices of $\mathcal{T}_1$ in $\widetilde{R}_{\mathcal{S}_0,\mathcal{S}_M,\mathcal{S}_\infty}$. If $t_2 \le 3$, then $Y = 0$.*

For example, Proposition 7.8 asserts that

$$Y_{ijk} = \Upsilon^a \Upsilon^b \Upsilon^c \Upsilon^d \widetilde{R}_{i\infty 0a,\infty bcj\infty d\infty k}|_{\rho=0,t=1} = 0.$$

Observe that the case $\mathcal{T}_1 = \emptyset$ follows from Proposition 7.6. In this case there are no contractions with $\operatorname{grad}\Upsilon$ and one need not restrict to $\{\rho = 0\}$.

Proof. The proof is by induction on r. By applying Proposition 6.1, we may as well assume $\mathcal{S}_0 = \emptyset$. By Proposition 7.6 we may assume $t_1 \ge 1$. Also it suffices to consider the case $t_2 = 3$, since any Y with $t_2 < 3$ may be viewed as a contraction against $\operatorname{grad}\Upsilon$'s of a Y with $t_2 = 3$ (Proposition 7.6 shows that Y vanishes unless $s_M = t_1 + t_2 \ge 4$, so there must exist enough indices contracted against $\operatorname{grad}\Upsilon$'s).

For $r = 0$, it need only be shown that $\Upsilon^l \widetilde{R}_{lijk}|_{\rho=0,t=1} = 0$. We have seen from (6.1) that $\widetilde{R}_{lijk}|_{\rho=0,t=1} = W_{lijk}$. The desired conclusion follows from the conformal transformation law $\widehat{C}_{ijk} = C_{ijk} - \Upsilon^l W_{lijk}$ for the Cotton tensor and the fact that the Cotton tensor vanishes for Einstein metrics.

Suppose the result is true up to $r - 1$. Construct a tensor $\widetilde{R}_{\emptyset,\mathcal{S}_M,\mathcal{S}_\infty}$ from $\widetilde{R}_{IJKL,M_1\cdots M_r}|_{\rho=0,t=1}$ as described above, where $s_M = t_1 + 3$ and $t_1 \ge 1$. We need to show that the contraction of $\operatorname{grad}\Upsilon$ into t_1 indices in $\mathcal{S}_M$ in the tensor $\widetilde{R}_{\emptyset,\mathcal{S}_M,\mathcal{S}_\infty}$ vanishes. Applying the Bianchi identity if necessary, we may assume that at least one of the indices of $\mathcal{S}_M$ is after the comma. By commuting derivatives, we want to arrange that $M_r \in \mathcal{S}_M$. Take the right-most element of $\mathcal{S}_M$ in the list $M_1 \cdots M_r$ and consider the effect of commuting it successively past each of the ∞'s to its right. According to the differentiated Ricci identity, the expression

$$\widetilde{R}_{IJKL,M_1\cdots M_s a\infty \underbrace{\infty\cdots\infty}_{k}} - \widetilde{R}_{IJKL,M_1\cdots M_s \infty a \underbrace{\infty\cdots\infty}_{k}}$$

is a sum of terms of the form

$$\widetilde{R}^{P}{}_{Na\infty,\underbrace{\infty\cdots\infty}_{l}} \widetilde{R}_{\dots,\,\dots\dots}$$

where $l \le k$, $N \in \{IJKLM_1 \cdots M_s\}$, and the second $\widetilde{R}$ has some list of indices including P but with N removed. Proposition 7.6 implies that

$\widetilde{R}^P{}_{Na\infty,\infty\cdots\infty} = 0$. Therefore the commutations introduce no new terms and we may assume that $M_r \in \mathcal{S}_M$. We write $M_r = a$.

Now write

$$\widetilde{R}_{IJKL,M_1\cdots M_{r-1}a} = \partial_a \widetilde{R}_{IJKL,M_1\cdots M_{r-1}} - \widetilde{\Gamma}^P_{aI}\widetilde{R}_{PJKL,M_1\cdots M_{r-1}} - \ldots - \widetilde{\Gamma}^P_{aM_{r-1}}\widetilde{R}_{IJKL,M_1\cdots P}. \tag{7.15}$$

Consider the term $\widetilde{\Gamma}^P_{aI}\widetilde{R}_{PJKL,M_1\cdots M_{r-1}}$. If $I = \infty$, (3.16) shows that only terms with $1 \le P \le n$ can contribute. Write $p = P$ and observe that (3.16) and (7.12) imply that $\widetilde{\Gamma}^p_{a\infty} = u\delta_a{}^p$ for some function $u(\rho)$, so that $\widetilde{\Gamma}^P_{aI}\widetilde{R}_{PJKL,M_1\cdots M_{r-1}} = u\widetilde{R}_{aJKL,M_1\cdots M_{r-1}}$. The induction hypothesis implies that this vanishes when restricted to $\rho = 0$ and contracted against t_1 factors of grad Υ. Similarly for the other terms in (7.15) involving Christoffel symbols $\widetilde{\Gamma}^P_{a\infty}$. Therefore we need consider only the terms involving Christoffel symbols of the form $\widetilde{\Gamma}^P_{ab}$ with $b \in \mathcal{S}_M \cap \{IJKLM_1\cdots M_{r-1}\}$. If, for example, $I = b \in \mathcal{S}_M$, such a term is of the form

$$\widetilde{\Gamma}^P_{ab}\widetilde{R}_{PJKL,M_1\cdots M_{r-1}} = \widetilde{\Gamma}^0_{ab}\widetilde{R}_{0JKL,M_1\cdots M_{r-1}} + \widetilde{\Gamma}^p_{ab}\widetilde{R}_{pJKL,M_1\cdots M_{r-1}} + \widetilde{\Gamma}^\infty_{ab}\widetilde{R}_{\infty JKL,M_1\cdots M_{r-1}}.$$

The number of elements of $\mathcal{T}_2$ in the list $JKLM_1\cdots M_{r-1}$ is of course at most 3. Therefore the induction hypothesis implies that upon setting $\rho = 0$ and contracting against the t_1 factors of grad Υ, the terms $\widetilde{\Gamma}^0_{ab}\widetilde{R}_{0JKL,M_1\cdots M_{r-1}}$ and $\widetilde{\Gamma}^\infty_{ab}\widetilde{R}_{\infty JKL,M_1\cdots M_{r-1}}$ both vanish. Similarly for the other Christoffel symbol terms corresponding to indices in $\mathcal{S}_M \cap \{IJKLM_1\cdots M_{r-1}\}$.

We have shown that we need only include terms on the right-hand side of (7.15) corresponding to indices in $\mathcal{S}_M \cap \{IJKLM_1\cdots M_{r-1}\}$ and for such terms need only sum over $p \in \{1,\ldots,n\}$. Since at $\rho = 0$ these ambient Christoffel symbols agree with the ones for the metric g on M, the right-hand side can be interpreted as $\nabla_a\widetilde{R}_{IJKL,M_1\cdots M_{r-1}}$, where ∇_a is the covariant derivative on M and the tensor $\widetilde{R}_{IJKL,M_1\cdots M_{r-1}}$ is to be interpreted as the rank $s_M - 1$ tensor on M determined by fixing the indices in $\mathcal{S}_\infty$ to be ∞ and letting the indices in $\mathcal{S}_M \setminus \{a\}$ vary between 1 and n. We must show that we get 0 if we contract t_1 factors of grad Υ into $\nabla_a\widetilde{R}_{IJKL,M_1\cdots M_{r-1}}$. Use the Leibnitz rule to factor the ∇_a outside the product of all the grad Υ terms with $\widetilde{R}_{IJKL,M_1\cdots M_{r-1}}$, at the expense of the sum of the terms where the derivative hits each one of the grad Υ factors in turn. By the induction hypothesis, the contraction of all the grad Υ factors (except Υ^a if it occurs) into $\widetilde{R}_{IJKL,M_1\cdots M_{r-1}}$ vanishes. To handle the terms where a derivative hits a grad Υ, recall the transformation law for the Schouten tensor:

$$\widehat{P}_{ij} = P_{ij} - \Upsilon_{ij} + \Upsilon_i\Upsilon_j - \frac{1}{2}\Upsilon^k\Upsilon_k g_{ij}.$$

We have $\widehat{P}_{ij} = \widehat{\lambda}\widehat{g}_{ij} = \widehat{\lambda}e^{2\Upsilon}g_{ij}$ and $P_{ij} = \lambda g_{ij}$. Therefore it follows that

$$\Upsilon_{ij} = \Upsilon_i\Upsilon_j + fg_{ij}$$

for some function f on M. Substituting this relation for the second derivatives of Υ which arise, one sees easily that the induction hypothesis implies that all the terms vanish, completing the induction. □

Proof of Proposition 7.5. We have already observed that for n odd, this follows from the uniqueness of the ambient metric up to diffeomorphism. So we can assume that $n \geq 4$ is even.

Recall that any pre-ambient metric for a given conformal class can be put into normal form corresponding to any choice of representative metric. So there is a diffeomorphism ψ such that to infinite order, $\psi^*\widetilde{g}$ is of the form (7.12) but with g_ρ replaced by some $\widehat{g}_\rho$ satisfying $\widehat{g}_0 = \widehat{g} = e^{2\Upsilon}g$. Moreover, the expansion of $\widehat{g}_\rho$ to order $n/2$ is uniquely determined by the Einstein condition $\mathrm{Ric}(\psi^*\widetilde{g}) = 0$ and the expansion to infinite order is determined once $\mathrm{tf}\left(\partial_\rho^{n/2}\widehat{g}_\rho|_{\rho=0}\right)$ is fixed. It follows that $\widehat{g}_\rho = (1+\widehat{\lambda}\rho)^2\widehat{g} + O(\rho^{n/2})$ and $\widehat{g}_\rho = (1+\widehat{\lambda}\rho)^2\widehat{g} + O(\rho^\infty)$ if and only if the curvature tensor of $\psi^*\widetilde{g}$ satisfies (7.14). We will show that the curvature tensor of $\psi^*\widetilde{g}$ satisfies (7.14), which will therefore prove the theorem.

The curvature tensor of $\psi^*\widetilde{g}$ is obtained from that of $\widetilde{g}$ by transforming it tensorially. The calculation of the Jacobian of ψ and this tensorial transformation are carried out in the proof of Proposition 6.5. The result is that all components of all covariant derivatives of curvature of $\psi^*\widetilde{g}$ are given by the right-hand side of (6.7). Thus (7.14) is equivalent to

$$\widetilde{R}_{ABCD,M_1\cdots M_{n/2-2}}|_{\rho=0}p^A{}_\infty p^B{}_i p^C{}_j p^D{}_\infty p^{M_1}{}_\infty \cdots p^{M_{n/2-2}}{}_\infty = 0,$$

where $p^A{}_I$ is given by (6.8). Expanding this out, one obtains a linear combination with smooth coefficients of contractions of some number (possibly 0) of factors of $\mathrm{grad}\,\Upsilon$ with tensors $\widetilde{R}_{\mathcal{S}_0,\mathcal{S}_M,\mathcal{S}_\infty}$ as considered in Proposition 7.8. Each contraction which occurs has $t_2 \leq 2$, so by Proposition 7.8, it vanishes.
□

The existence of a unique normalized infinite order ambient metric up to diffeomorphism in the locally conformally flat and conformally Einstein cases has as a consequence that other invariant objects exist for such structures which do not exist for general conformal structures in even dimensions. An example is the family of "conformally invariant powers of the Laplacian" constructed in [GJMS]. In [GJMS] it was shown that if $n \geq 3$ is odd and $k \in \mathbb{N}$, then there is a natural scalar differential operator with principal part $(-\Delta)^k$ depending on a metric g, which defines a conformally invariant operator $P_{2k} : \mathcal{E}(-n/2+k) \to \mathcal{E}(-n/2-k)$, where $\mathcal{E}(w)$ denotes the space of conformal densities of weight w. (Recall that our convention is $\Delta = \nabla^i \nabla_i$.) When $n \geq 4$ is even, the same result holds with the restriction $k \leq n/2$, and it was shown in [GoH], following the special case $k = 3$, $n = 4$ established in [Gr1], that no such natural operator exists for $k > n/2$ for general conformal structures. The conformal invariance of the operator produced by the algorithm in [GJMS] depends only on the existence and invariance of the ambient metric. So it follows that when n is even and $(M, [g])$ is locally conformally flat or conformally Einstein, there exists for all $k \geq 1$ an operator $P_{2k} : \mathcal{E}(-n/2 + k) \to \mathcal{E}(-n/2 - k)$ with principal part $(-\Delta)^k$ determined solely by the conformal structure.

One can write the operator P_{2k} explicitly for an Einstein metric g. In [GJMS], the operator P_{2k} arose as the obstruction to "harmonic extension" of the density. We briefly review the construction. A density of weight w can be viewed as a homogeneous function of degree w on $\mathcal{G}$, and one asks for an extension to a homogeneous function of degree w on $\mathcal{G} \times \mathbb{R}$ which solves $\widetilde{\Delta} u = 0$ to high order along $\mathcal{G} \times \{0\}$, where $\widetilde{\Delta}$ denotes the Laplacian with respect to an ambient metric $\widetilde{g}$. In the identification $\mathcal{G} \times \mathbb{R} \cong \mathbb{R}_+ \times M \times \mathbb{R}$ induced by a choice of metric g in the conformal class, we can write $u = t^w f$, where $f = f(x, \rho)$ is a function whose restriction to $\rho = 0$ is the initial density on M. The equation $\widetilde{\Delta} u = 0$ is equivalent to the following equation on f:

$$-2\rho f'' + (2w + n - 2 - \rho g^{ij} g'_{ij}) f' + \Delta_\rho f + \tfrac{1}{2} w g^{ij} g'_{ij} f = 0, \tag{7.16}$$

where $'$ denotes ∂_ρ, $g_{ij}(x, \rho)$ is as in (3.14), and Δ_ρ denotes the Laplacian with respect to $g_{ij}(x, \rho)$ with ρ fixed. For $w = -n/2 + k$, the derivatives $\partial_\rho^m f|_{\rho=0}$ for $1 \leq m \leq k - 1$ are determined successively by differentiating (7.16) with respect to ρ at $\rho = 0$, but the expression obtained by differentiating the left-hand side of (7.16) $k - 1$ times depends only on the previously determined derivatives so defines an obstruction to harmonic extension which can be expressed modulo a normalizing constant as the invariant operator P_{2k} applied to $f(x, 0)$. For general metrics one must require $k \leq n/2$ if n is even because g_{ij} is then only determined to order $n/2$, but this is not necessary if the initial metric is Einstein.

Suppose now that the initial metric g is Einstein with $\mathrm{Ric}(g) = 2\lambda(n-1)g$ as above. Then $g_{ij}(x, \rho) = (1 + \lambda\rho)^2 g_{ij}(x, 0)$, so for $w = -n/2 + k$, (7.16) becomes

$$-2\rho f'' + \left(2k - 2 - \frac{2n\lambda\rho}{1 + \lambda\rho}\right) f' + \frac{1}{(1 + \lambda\rho)^2} \Delta f + \frac{\lambda n(-n/2 + k)}{1 + \lambda\rho} f = 0, \tag{7.17}$$

where $\Delta = \Delta_0$ denotes the Laplacian with respect to the initial metric g. It is clear from the form of this equation and from the inductive procedure above that in this case the operator P_{2k} takes the form $P_{2k} = p_{2k}(\Delta, \lambda)$, where p_{2k} is a polynomial in two real variables depending on n and k as parameters.

Moreover, rescaling ρ in (7.17) and using the fact that P_{2k} has leading coefficient $(-\Delta)^k$ shows that $p_{2k}(\Delta,\lambda) = \lambda^k p_{2k}(\Delta/\lambda, 1)$. Thus P_{2k} for Einstein metrics is determined by the polynomial $p_{2k}(\Delta, 1)$. In [Br], Branson showed that on S^n, any differential operator P_{2k} which defines an equivariant map $: \mathcal{E}(-n/2+k) \to \mathcal{E}(-n/2-k)$ on densities viewed as sections of homogeneous vector bundles for the conformal group is necessarily a multiple of

$$\prod_{j=1}^{k} (-\Delta + c_j), \qquad c_j = \left(\tfrac{n}{2} + j - 1\right)\left(\tfrac{n}{2} - j\right)$$

when written with respect to the metric of constant sectional curvature 1. The operator P_{2k} has this invariance property as a consequence of its naturality and invariance under conformal rescaling, so this determines the polynomial p_{2k}. Thus we deduce

Proposition 7.9. *If $n \geq 3$ and if g is Einstein with* $\mathrm{Ric}(g) = 2\lambda(n-1)g$, *then the operators P_{2k} are given for $k \geq 1$ by*

$$P_{2k} = \prod_{j=1}^{k} (-\Delta + 2\lambda c_j), \qquad c_j = \left(\tfrac{n}{2} + j - 1\right)\left(\tfrac{n}{2} - j\right).$$

The above argument showing that the operators produced by the GJMS algorithm for the ambient metric (7.12) of an Einstein metric are given by the same formula as on the sphere was presented by the second author at the 2003 AIM Workshop on Conformal Structure in Geometry, Analysis, and Physics in response to a question of Alice Chang (see [BEGW]). One can also argue via Proposition 3.5 if n is odd or if n is even and $k \leq n/2$. In these cases, Proposition 3.5 implies that the coefficients of the P_{2k} can be written solely in terms of the Ricci curvature and its covariant derivatives. For an Einstein metric, the covariant derivatives of Ricci vanish, so the formula necessarily reduces to a universal formula involving only Δ and λ. Another treatment of these results has been given by Gover in [Go2] using tractors.

Proposition 7.9 can also be derived without reference to Branson's formula on the sphere directly from the recursion relation obtained by successively differentiating (7.17). Upon setting $\lambda = 1$, multiplying (7.17) by $(1+\rho)^2$, and differentiating m times, one obtains at $\rho = 0$

$$\begin{aligned} 2(k-m-1)f^{(m+1)} + \left[\Delta + 4m(k-m-n/2) + n(k-n/2)\right] f^{(m)} \\ + m\left[2(m-1)(k-m+1-n) + n(k-n/2)\right] f^{(m-1)} = 0. \end{aligned}$$

Set $(a)_0 = 1$ and $(a)_m = a(a+1)\dots(a+m-1)$ for $a \in \mathbb{C}$, $m \in \mathbb{N}$. It follows that

$$(k-m)_m f^{(m)}(x,0) = q_m(y) f(x,0) \qquad 0 \leq m \leq k-1$$

and that

$$p_{2k}(\Delta, 1) = 2^k q_k(y), \tag{7.18}$$

where $y = -\frac{1}{2}[\Delta - n(n/2-1)]$ and the $q_m(y)$ are the polynomials of one variable determined by $q_{-1} = 0$, $q_0 = 1$ and the recursion relation

$$\begin{aligned} q_{m+1} = &[y - 2m(k-m-n/2) - n(k-1)/2]\, q_m \\ &- m(m-k)(m-1+n/2)(m-1+n/2-k)q_{m-1} \end{aligned} \tag{7.19}$$

for $m \geq 0$.

Up to a normalizing factor, the polynomials q_m are a case of the dual Hahn polynomials, a family of discrete orthogonal polynomials which may be expressed explicitly in terms of ${}_3F_2$ hypergeometric functions (see [KM], [NSU] and the appendix of [AW]). This leads to the following formula for the q_m:

$$q_m(y) = \sum_{l=0}^{m} (-1)^{m-l}(n/2+l)_{m-l}(k-m)_{m-l}\binom{m}{l}\prod_{j=1}^{l}[y - j(j-1)],$$

where for $l = 0$ the empty product is interpreted as 1. Indeed, it is not difficult to verify directly that q_m defined by this formula satisfies (7.19). Taking $m = k$ gives

$$q_k(y) = \prod_{j=1}^{k}[y - j(j-1)],$$

which via (7.18) gives Proposition 7.9.

We remark that Proposition 7.9 gives a formula for Branson's Q-curvature for an Einstein metric. From Branson's original definition in [Br], one obtains immediately from Proposition 7.9 that if n is even and $\operatorname{Ric}(g) = 2\lambda(n-1)g$, then

$$Q(g) = (2\lambda)^{n/2}(n-1)\prod_{j=1}^{n/2-1}(\tfrac{n}{2}+j-1)(\tfrac{n}{2}-j).$$

Chapter Eight

Jet Isomorphism

A fundamental result in Riemannian geometry is the jet isomorphism theorem which asserts that at the origin in geodesic normal coordinates, the full Taylor expansion of the metric may be recovered from the iterated covariant derivatives of curvature. As a consequence, one deduces that any local invariant of Riemannian metrics has a universal expression in terms of the curvature tensor and its covariant derivatives. Geodesic normal coordinates are determined up to the orthogonal group, so problems involving local invariants are reduced to purely algebraic questions concerning invariants of the orthogonal group on tensors.

Our goal in this chapter is to prove an analogous jet isomorphism theorem for conformal geometry. By making conformal changes, the Taylor expansion of a metric in geodesic normal coordinates can be further simplified, resulting in a "conformal normal form" for metrics about a point. The jet isomorphism theorem states that the map from the Taylor coefficients of metrics in conformal normal form to the space of all conformal curvature tensors, realized in terms of covariant derivatives of ambient curvature, is an isomorphism. If n is even, the theorem holds only up to a finite order. In the conformal case, the role of the orthogonal group in Riemannian geometry is played by a parabolic subgroup of the conformal group. We assume throughout this chapter and the next that $n \geq 3$.

We begin by reviewing the Riemannian theorem in the form we will use it. Fix a reference quadratic form h_{ij} of signature (p, q) on $\mathbb{R}^n$. In the positive definite case one typically chooses $h_{ij} = \delta_{ij}$. The background coordinates are geodesic normal coordinates for a metric g_{ij} defined near the origin in $\mathbb{R}^n$ if and only if $g_{ij}(0) = h_{ij}$ and the radial vector field ∂_r satisfies $\nabla_{\partial_r}\partial_r = 0$, which is equivalent to $\Gamma^i_{jk}x^j x^k = 0$ or

$$2\partial_k g_{ij}x^j x^k = \partial_i g_{jk}x^j x^k. \tag{8.1}$$

It is easily seen that the coordinates are normal if and only if

$$g_{ij}x^j = h_{ij}x^j. \tag{8.2}$$

In fact, set $F_i = g_{ij}x^j - h_{ij}x^j$ and observe that

$$\partial_k g_{ij}x^j x^k = x^j\partial_j F_i - F_i, \qquad \partial_i g_{jk}x^j x^k = \partial_i(x^j F_j) - 2F_i. \tag{8.3}$$

Clearly $F_i = 0$ implies that (8.1) holds so that the coordinates are normal. Conversely, if the coordinates are normal, then the fact that g and ∂_r are parallel along radial lines implies $x^j F_j = 0$. Substituting this and (8.3) into (8.1) gives $x^j \partial_j F_i = 0$, which together with the fact that F_i is smooth and vanishes at the origin implies $F_i = 0$.

We are interested in the space of infinite order jets of metrics at the origin. Taylor expanding shows that (8.2) holds to infinite order if and only if $g_{ij}(0) = h_{ij}$ and the derivatives of g_{ij} satisfy $\partial_{(k_1} \dots \partial_{k_r} g_{i)j}(0) = 0$. Since a 3-tensor symmetric in 2 indices and skew in 2 indices must vanish, this implies in particular that all first derivatives of g vanish at the origin. So we define the space $\mathcal{N}$ of jets of metrics in geodesic normal coordinates as follows.

Definition 8.1. The space $\mathcal{N}$ is the set of lists $(g^{(2)}, g^{(3)}, \dots)$, where for each r, $g^{(r)} \in \bigodot^2 \mathbb{R}^{n*} \otimes \bigodot^r \mathbb{R}^{n*}$ satisfies $g^{(r)}_{i(j,k_1 \cdots k_r)} = 0$. Here the comma just serves to separate the first two indices. For $N \geq 2$, $\mathcal{N}^N$ will denote the set of truncated lists $(g^{(2)}, g^{(3)}, \dots, g^{(N)})$ with the same conditions on the $g^{(r)}$.

To a metric in geodesic normal coordinates near the origin, we associate the element of $\mathcal{N}$ given by $g^{(r)}_{ij,k_1 \cdots k_r} = \partial_{k_1} \dots \partial_{k_r} g_{ij}(0)$ for $r \geq 2$. Conversely, to an element of $\mathcal{N}$ we associate the metric determined to infinite order by these relations together with $g_{ij}(0) = h_{ij}$ and $\partial_k g_{ij}(0) = 0$. In the following, we typically identify the element of $\mathcal{N}$ and the jet of the metric g.

Definition 8.2. The space $\mathcal{R}$ is the set of lists $(R^{(0)}, R^{(1)}, \dots)$, where for each r, $R^{(r)} \in \bigwedge^2 \mathbb{R}^{n*} \otimes \bigwedge^2 \mathbb{R}^{n*} \otimes \bigotimes^r \mathbb{R}^{n*}$, and the usual identities satisfied by covariant derivatives of curvature hold:

(1) $R_{i[jkl],m_1 \cdots m_r} = 0$;

(2) $R_{ij[kl,m_1]m_2 \cdots m_r} = 0$;

(3) $R_{ijkl,m_1 \cdots [m_{s-1} m_s] \cdots m_r} = Q^{(s)}_{ijklm_1 \cdots m_r}$.
Here $Q^{(s)}_{ijklm_1 \cdots m_r}$ denotes the quadratic polynomial in the $R^{(r')}$ with $r' \leq r - 2$ which one obtains by covariantly differentiating the usual Ricci identity for commuting covariant derivatives, expanding the differentiations using the Leibnitz rule, and then setting equal to h the metric which contracts the two factors in each term.

For $N \geq 0$, $\mathcal{R}^N$ will denote the set of truncated lists $(R^{(0)}, R^{(1)}, \dots, R^{(N)})$ with the same conditions on the $R^{(r)}$.

There is a natural map $\mathcal{N} \to \mathcal{R}$ induced by evaluation of the covariant derivatives of curvature of a metric, which is polynomial in the sense that the corresponding truncated maps $\mathcal{N}^{N+2} \to \mathcal{R}^N$ are polynomial. We will say that a map on a subset of a finite-dimensional vector space is polynomial if it is the restriction of a polynomial map defined on the whole space, and a map between subsets of finite-dimensional vector spaces is a polynomial equivalence if it is a bijective polynomial map whose inverse is also polynomial. The jet isomorphism theorem for pseudo-Riemannian geometry is the following.

Theorem 8.3. *The map $\mathcal{N} \to \mathcal{R}$ is bijective and for each $N \geq 0$ the truncated map $\mathcal{N}^{N+2} \to \mathcal{R}^N$ is a polynomial equivalence.*

There are two parts to the proof. One is a linearization argument showing that it suffices to show that the linearized map is an isomorphism. The other is the observation that the linearized map is the direct sum over r of isomorphisms between two realizations according to different Young projectors of specific irreducible representations of $GL(n, \mathbb{R})$. The reduction to the linearization can be carried out in different ways; one is to argue as we do below in the conformal case. The analysis of the linearized map is contained in [E2]. See also [ABP] for a different construction of a left inverse of the map $\mathcal{N} \to \mathcal{R}$.

We now consider the further freedom allowed by the possibility of making conformal changes to g.

Proposition 8.4. *Let g be a metric on a manifold M and let $p \in M$. Given $\Omega_0 \in C^\infty(M)$ with $\Omega_0(p) > 0$, there is $\Omega \in C^\infty(M)$, uniquely determined to infinite order at p, such that $\Omega - \Omega_0$ vanishes to second order at p and $\mathrm{Sym}(\nabla^r \mathrm{Ric})(\Omega^2 g)(p) = 0$ for all $r \geq 0$. Here $\mathrm{Sym}(\nabla^r \mathrm{Ric})(\Omega^2 g)$ denotes the full symmetrization of the rank $r+2$ tensor $(\nabla^r \mathrm{Ric})(\Omega^2 g)$.*

Proof. Write $\Omega = e^\Upsilon$ and set $\widehat{g} = e^{2\Upsilon} g$. We are given $\Upsilon(p)$ and $d\Upsilon(p)$. Recall the transformation law for conformal change of Ricci curvature:

$$\widehat{R}_{ij} = R_{ij} - (n-2)\Upsilon_{ij} - \Upsilon_k{}^k g_{ij} + (n-2)(\Upsilon_i \Upsilon_j - \Upsilon_k \Upsilon^k g_{ij}).$$

Differentiating and conformally transforming the covariant derivative results in

$$\begin{aligned}(\widehat{\nabla^r \mathrm{Ric}})_{ij,m_1\cdots m_r} =& (\nabla^r \mathrm{Ric})_{ij,m_1\cdots m_r} - (n-2)\Upsilon_{ijm_1\cdots m_r} \\ &- \Upsilon_k{}^k{}_{m_1\cdots m_r} g_{ij} + \mathrm{lots},\end{aligned}$$

where lots denotes terms involving at most $r+1$ derivatives of Υ. We may replace covariant derivatives of Υ by coordinate derivatives on the right-hand side of this formula. Then symmetrizing gives

$$\mathrm{Sym}\,(\widehat{\nabla^r \mathrm{Ric}}) = \mathrm{Sym}(\nabla^r \mathrm{Ric}) - (n-2)\partial^{r+2}\Upsilon - \mathrm{Sym}(\mathrm{tr}(\partial^{r+2}\Upsilon) g) + \mathrm{lots}.$$

If $c > 0$, the map $s \to s + c\,\mathrm{Sym}(\mathrm{tr}(s)g)$ is invertible on symmetric $(r+2)$-tensors. Therefore, for any $r \geq 0$, one can uniquely determine $\partial^{r+2}\Upsilon(p)$ to make $\mathrm{Sym}\,(\widehat{\nabla^r \mathrm{Ric}})(p) = 0$. The result follows upon iterating and applying Borel's Lemma. □

We remark that there are other natural choices for normalizations of the conformal factor. For example, one such is that the symmetrized covariant derivatives of the tensor P_{ij} vanish at p, where P_{ij} is given by (3.7). Another is that in normal coordinates, $\det g_{ij} - 1$ vanishes to infinite order at p (see [LP]).

Definition 8.5. The space $\mathcal{N}_c \subset \mathcal{N}$ of jets of metrics in conformal normal form is the subset of $\mathcal{N}$ consisting of jets of metrics in geodesic normal coordinates for which $\mathrm{Sym}(\nabla^r \mathrm{Ric})(g)(0) = 0$ for all $r \geq 0$. For $N \geq 2$, $\mathcal{N}_c^N$ will denote the subset of $\mathcal{N}^N$ obtained by requiring that this relation hold for $0 \leq r \leq N-2$.

Later we will need the following consequence of the proof of Proposition 8.4.

Lemma 8.6. *For each $N \geq 3$, there is a polynomial map $\eta^N : \mathcal{N}_c^{N-1} \to \bigodot^2 \mathbb{R}^{n*} \otimes \bigodot^N \mathbb{R}^{n*}$ with zero constant and linear terms, such that the map $g \to (g, \eta^N(g))$ maps $\mathcal{N}_c^{N-1} \to \mathcal{N}_c^N$. Here $g = (g^{(2)}, g^{(3)}, \ldots, g^{(N-1)}) \in \mathcal{N}_c^{N-1}$.*

Proof. View g as the metric on a neighborhood of $0 \in \mathbb{R}^n$ given by the prescribed finite Taylor expansion of order $N-1$. Then the components of $\mathrm{Sym}(\nabla^{N-2} \mathrm{Ric})(g)(0)$ are polynomials in the $g^{(r)}$ with no constant or linear terms. According to the proof of Proposition 8.4, there is a function Ω of the form $\Omega = 1 + p_N$ with p_N a homogeneous polynomial of degree N whose coefficients are polynomials in the $g^{(r)}$ with no constant or linear terms, such that $\mathrm{Sym}(\nabla^{N-2} \mathrm{Ric})(\Omega^2 g)(0) = 0$. Now $\Omega^2 g$ is in geodesic normal coordinates to order $N-1$ but not necessarily N. However by the construction of geodesic normal coordinates, there is a diffeomorphism $\psi = I + q_{N+1}$, where q_{N+1} is a vector-valued homogeneous polynomial of order $N+1$ whose coefficients are linear in the order N Taylor coefficient of $\Omega^2 g$, such that $\psi^*(\Omega^2 g)$ is in geodesic normal coordinates to order N. Since the condition $\mathrm{Sym}(\nabla^r \mathrm{Ric})(\Omega^2 g)(0) = 0$ is invariant under diffeomorphisms, η^N defined to be the order N Taylor coefficient of $\psi^*(\Omega^2 g)$ has the required properties. □

If g_1 and g_2 are jets of metrics of signature (p,q) at the origin in $\mathbb{R}^n$, we say that g_1 and g_2 are equivalent if there is a local diffeomorphism ψ defined near 0 satisfying $\psi(0) = 0$, and a positive smooth function Ω defined near 0, so that $g_2 = \psi^*(\Omega^2 g_1)$ to infinite order. It is clear from Proposition 8.4

and the existence of geodesic normal coordinates that any jet of a metric is equivalent to one in $\mathcal{N}_c$. In choosing Ω we have the freedom of $\mathbb{R}_+ \times \mathbb{R}^n$, and in choosing ψ a freedom of $O(p,q)$. We next describe how these freedoms can be realized as an action on $\mathcal{N}_c$ of a subgroup of the conformal group.

Recall that h_{ij} is our fixed background quadratic form of signature (p,q) on $\mathbb{R}^n$. Define a quadratic form $\widetilde{h}_{IJ}$ on $\mathbb{R}^{n+2}$ by

$$\widetilde{h}_{IJ} = \begin{pmatrix} 0 & 0 & 1 \\ 0 & h_{ij} & 0 \\ 1 & 0 & 0 \end{pmatrix}$$

and the quadric $\mathcal{Q} = \{[x^I] : \widetilde{h}_{IJ}x^I x^J = 0\} \subset \mathbb{P}^{n+1}$. The metric $\widetilde{h}_{IJ}dx^I dx^J$ on $\mathbb{R}^{n+2}$ induces a conformal structure of signature (p,q) on $\mathcal{Q}$. The standard action of the orthogonal group $O(\widetilde{h})$ on $\mathbb{R}^{n+2}$ induces an action on $\mathcal{Q}$ by conformal transformations and the adjoint group $O(\widetilde{h})/\{\pm I\}$ can be identified with the conformal group of all conformal transformations of $\mathcal{Q}$. Let $e_0 = \begin{pmatrix} 1 \\ 0 \\ 0 \end{pmatrix} \in \mathbb{R}^{n+2}$ and let P be the image in $O(\widetilde{h})/\{\pm I\}$ of the isotropy group $\{A \in O(\widetilde{h}) : Ae_0 = ae_0, a \in \mathbb{R}\}$ of $[e_0]$. It is clear that each element of P is represented by exactly one A for which $a > 0$, so we make the identification $P = \{p \in O(\widetilde{h}) : pe_0 = ae_0, a > 0\}$. The first column of $p \in P$ is $\begin{pmatrix} a \\ 0 \\ 0 \end{pmatrix}$; combining this with the fact that $p \in O(\widetilde{h})$, one finds that

$$P = \left\{ \begin{pmatrix} a & b_j & c \\ 0 & m^i{}_j & d^i \\ 0 & 0 & a^{-1} \end{pmatrix} : a > 0,\ m^i{}_j \in O(h),\ c = -\frac{1}{2a} b_j b^j,\ d^i = -\frac{1}{a} m^{ij} b_j \right\}$$

where h_{ij} is used to raise and lower lowercase indices. It is evident that $P = \mathbb{R}_+ \cdot \mathbb{R}^n \cdot O(h)$, where the subgroups $\mathbb{R}_+$, $\mathbb{R}^n$, $O(h)$ arise by varying a, b_j, $m^i{}_j$, resp.

The intersection of $\mathcal{Q}$ with the cell $\{[x^I] : x^0 \neq 0\}$ can be identified with $\mathbb{R}^n$ via $\mathbb{R}^n \ni x^i \to \begin{bmatrix} 1 \\ x^i \\ -\frac{1}{2}|x|^2 \end{bmatrix} \in \mathcal{Q}$, where $|x|^2 = h_{ij}x^i x^j$. In this identification, the conformal structure is represented by the metric $h_{ij}dx^i dx^j$ on $\mathbb{R}^n$. If $p \in P$ is as above, the conformal transformation determined by p is realized as

$$(\varphi_p(x))^i = \frac{m^i{}_j x^j - \frac{1}{2}|x|^2 d^i}{a + b_j x^j - \frac{1}{2}c|x|^2}$$

and one has $\varphi_p^* h = \Omega_p^2 h$ for

$$\Omega_p = (a + b_j x^j - \tfrac{1}{2}c|x|^2)^{-1}.$$

This motivates the following definition of an action of P on $\mathcal{N}_c$. Given $p \in P$ as above and $g \in \mathcal{N}_c$, by Proposition 8.4 there is a positive smooth function Ω uniquely determined to infinite order at 0 so that Ω agrees with Ω_p to second order and such that $\mathrm{Sym}(\nabla^r \mathrm{Ric})(\Omega^2 g)(0) = 0$ for all $r \geq 0$. Now $(\Omega^2 g)(0) = a^{-2}h$, so by the construction of geodesic normal coordinates, there is a diffeomorphism φ, uniquely determined to infinite order at 0, so that $\varphi(0) = 0$, $\varphi'(0) = a^{-1}m$, and such that $(\varphi^{-1})^*(\Omega^2 g)$ is in geodesic normal coordinates to infinite order. We define $p.g = (\varphi^{-1})^*(\Omega^2 g)$. It is clear by construction of φ that $p.g \in \mathcal{N}$, and since the condition of vanishing of the symmetrized covariant derivatives of Ricci curvature is diffeomorphism-invariant, it follows by construction of Ω that $p.g \in \mathcal{N}_c$. It is straightforward to check that this defines a left action of P on $\mathcal{N}_c$. Note that if $g = h$, then $\Omega = \Omega_p$ and $\varphi = \varphi_p$, so that h is a fixed point of the action. A moment's thought shows that $(p.g)^{(r)}$ depends only on $g^{(s)}$ for $s \leq r$. Therefore for each $N \geq 2$, there is an induced action on $\mathcal{N}_c^N$.

It is clear from the construction of the action that $p.g$ is equivalent to g for all $g \in \mathcal{N}_c$ and $p \in P$. In fact, the P-orbits are exactly the equivalence classes.

Proposition 8.7. *The orbits of the P-action on $\mathcal{N}_c$ are precisely the equivalence classes of jets of metrics in $\mathcal{N}_c$ under diffeomorphism and conformal change.*

Proof. It remains to show that equivalent jets of metrics in $\mathcal{N}_c$ are in the same P-orbit. Suppose that $g_1, g_2 \in \mathcal{N}_c$ are equivalent. Then we can write $g_2 = (\varphi^{-1})^*(\Omega^2 g_1)$ to infinite order for a diffeomorphism φ with $\varphi(0) = 0$ and a positive smooth Ω. We can uniquely choose the parameters a and b of $p \in P$ so that $\Omega - \Omega_p$ vanishes to second order. Since g_1 and g_2 both equal h at 0, it follows that $a\varphi'(0) \in O(h)$, so that we can write $\varphi'(0) = a^{-1}m$ with $m \in O(h)$. Together with the already determined parameters a and b, this choice of m uniquely determines a $p \in P$. Since $g_2 \in \mathcal{N}_c$, all symmetrized covariant derivatives of Ricci curvature of $\Omega^2 g_1$ vanish at the origin, so Ω must be the conformal factor determined when constructing the action of p on g_1. And since g_2 is in geodesic normal coordinates, φ must be the correct diffeomorphism, so that $g_2 = p.g_1$. $\square$

It is straightforward to calculate from the definition the action on $\mathcal{N}_c$ of the $\mathbb{R}_+$ and $O(h)$ subgroups of P. If we denote by p_a the element of P obtained by taking $b = 0$ and $m = I$ and by p_m the element given by $a = 1$ and $b = 0$, then one finds that p_a acts by multiplying $g^{(r)}$ by a^r, and p_m acts by transforming each $g^{(r)}$ as an element of $\bigotimes^{r+2} \mathbb{R}^{n*}$, where $\mathbb{R}^n$ denotes the standard defining representation of $O(h)$.

The problem of understanding local invariants of metrics under diffeomorphism and conformal change reduces to understanding this action of P on $\mathcal{N}_c$. However, it is very difficult to analyze or even concretely exhibit the action of the $\mathbb{R}^n$ part of P directly from the definition. The ambient curvature tensors enable the reformulation of the action in terms of standard tensor representations of P.

Propositions 6.1 and 6.4 show that the covariant derivatives of curvature of an ambient metric $\widetilde{g}$ satisfy relations arising from the homogeneity and Ricci-flatness of the metric. These conditions suggest the following definition.

Definition 8.8. The space $\widetilde{\mathcal{R}}$ is the set of lists $(\widetilde{R}^{(0)}, \widetilde{R}^{(1)}, \ldots)$, such that $\widetilde{R}^{(r)} \in \bigwedge^2 \mathbb{R}^{n+2*} \otimes \bigwedge^2 \mathbb{R}^{n+2*} \otimes \bigotimes^r \mathbb{R}^{n+2*}$, and such that the following relations hold:

(1) $\widetilde{R}_{I[JKL],M_1\cdots M_r} = 0$;

(2) $\widetilde{R}_{IJ[KL,M_1]M_2\cdots M_r} = 0$;

(3) $\widetilde{h}^{IK}\widetilde{R}_{IJKL,M_1\cdots M_r} = 0$;

(4) $\widetilde{R}_{IJKL,M_1\cdots[M_{s-1}M_s]\cdots M_r} = \widetilde{Q}^{(s)}_{IJKLM_1\cdots M_r}$;

(5) $\widetilde{R}_{IJK0,M_1\cdots M_r} = -\sum_{s=1}^r \widetilde{R}_{IJKM_s,M_1\cdots\widehat{M_s}\cdots M_r}$;

(6) $\widetilde{R}_{IJKL,M_1\cdots M_s 0 M_{s+1}\cdots M_r} =$
$$-(s+2)\widetilde{R}_{IJKL,M_1\cdots M_r} - \sum_{t=s+1}^r \widetilde{R}_{IJKL,M_1\cdots M_s M_t M_{s+1}\cdots\widehat{M_t}\cdots M_r}.$$

Here, as in Definition 8.2, $\widetilde{Q}^{(s)}_{IJKLM_1\cdots M_r}$ denotes the quadratic polynomial in the components of $\widetilde{R}^{(r')}$ for $r' \le r-2$ which one obtains by differentiating the Ricci identity for commuting covariant derivatives, expanding the differentiations using the Leibnitz rule, and then setting equal to $\widetilde{h}$ the metric which contracts the two factors in each term. Condition (5) in case $r = 0$ is interpreted as $\widetilde{R}_{IJK0} = 0$.

We remark, as we did in the proof of Proposition 6.1, that condition (6) is superfluous: it is a consequence of (2), (4), (5). But we will not use this fact. It will be convenient also to introduce the vector space $\widetilde{\mathcal{T}}$ consisting of the set of lists $(\widetilde{R}^{(0)}, \widetilde{R}^{(1)}, \ldots)$ with $\widetilde{R}^{(r)} \in \bigwedge^2 \mathbb{R}^{n+2*} \otimes \bigwedge^2 \mathbb{R}^{n+2*} \otimes \bigotimes^r \mathbb{R}^{n+2*}$ such that (1)–(3), (5), (6) hold.

We prepare to define truncated spaces $\widetilde{\mathcal{R}}^N$ for $\widetilde{\mathcal{R}}$. Recall the notion of strength from Definition 6.3. Note that it is clear that for each of the relations (1)–(6) except (4) in Definition 8.8, all components $\widetilde{R}_{IJKL,M_1\cdots M_r}$ which occur in the relation have the same value for the strength of the index list of the component. For $N \geq 0$, define the following vector spaces of lists of components of tensors. We denote by $\mathcal{M} = M_1 \cdots M_r$ a list of indices of length $|\mathcal{M}| = r$, and the conditions (1)–(6) refer to Definition 8.8.

$$\widetilde{\mathcal{T}}^N = \{(\widetilde{R}_{IJKL,\mathcal{M}})_{|\mathcal{M}|\geq 0,\ \|IJKL\mathcal{M}\|\leq N+4} : \text{ (1)–(3), (5), (6) hold}\}$$

$$\widetilde{\tau}^N = \{(\widetilde{R}_{IJKL,\mathcal{M}})_{|\mathcal{M}|\geq 0,\ \|IJKL\mathcal{M}\|= N+4} : \text{ (1)–(3), (5), (6) hold}\}$$

If $\widetilde{R}_{IJKL,M_1\cdots M_r}$ is a component appearing in an element of $\widetilde{\mathcal{T}}^N$ and $r > N$, then at least one of the indices $IJKLM_1 \cdots M_r$ must be 0. Therefore by (5) and (6), $\widetilde{R}_{IJKL,M_1\cdots M_r}$ can be written as a linear combination of components with r replaced by $r-1$. It follows that $\widetilde{\mathcal{T}}^N$ and $\widetilde{\tau}^N$ are finite-dimensional. Since (1)–(3), (5) and (6) imply that $\widetilde{R}_{IJKL,M_1\cdots M_r} = 0$ if $\|IJKLM_1 \cdots M_r\| \leq 3$, we have $\widetilde{\mathcal{T}}^N = \bigoplus_{M=0}^N \widetilde{\tau}^M$ and $\widetilde{\mathcal{T}} = \prod_{M=0}^\infty \widetilde{\tau}^M$.

As for (4), a typical term in $\widetilde{Q}^{(s)}_{IJKLM_1\cdots M_r}$ is

$$\widetilde{h}^{AB}\widetilde{R}_{AIM_{s-1}M_s,\mathcal{M}'}\widetilde{R}_{BJKL,M_1\cdots M_{s-2}\mathcal{M}''},$$

where $\mathcal{M}'$ and $\mathcal{M}''$ are lists of indices such that $\mathcal{M}'\mathcal{M}''$ is a rearrangement of $M_{s+1} \cdots M_r$. In order that $\widetilde{h}^{AB} \neq 0$, it must be that $\|AB\| = 2$. Therefore $\|AIM_{s-1}M_s\mathcal{M}'\| + \|BJKLM_1 \cdots M_{s-2}\mathcal{M}''\| = \|IJKLM_1 \cdots M_r\| + 2$. This implies that if either $\|AIM_{s-1}M_s\mathcal{M}'\|$ or $\|BJKLM_1 \cdots M_{s-2}\mathcal{M}''\|$ is greater than or equal to $\|IJKLM_1 \cdots M_r\|$, then the other is less than or equal to 2. The same reasoning applies to all terms in $\widetilde{Q}^{(s)}_{IJKLM_1\cdots M_r}$. Since (1)–(3), (5) and (6) imply that a component of an $\widetilde{R}^{(t)}$ vanishes if its strength is at most 3, it follows that any component of an $\widetilde{R}^{(t)}$ which occurs in $\widetilde{Q}^{(s)}_{IJKLM_1\cdots M_r}$ with a nonzero coefficient must have strength strictly less than $\|IJKLM_1 \cdots M_r\|$. Hence we will regard (4) as a relation involving components with indices of strength at most $\|IJKLM_1 \cdots M_r\|$, and the quadratic terms in (4) only involve components of strength less than that of the linear terms. With this understanding, we can now define

$$\widetilde{\mathcal{R}}^N = \{(\widetilde{R}_{IJKL,M_1\cdots M_r}) \in \widetilde{\mathcal{T}}^N : \text{ (4) of Definition 8.8 holds}\}.$$

We will also need the corresponding linearized spaces, in which the term $\widetilde{Q}^{(s)}_{IJKLM_1\cdots M_r}$ in (4) is replaced by 0. Define vector spaces

$$T\widetilde{\mathcal{R}} = \{(\widetilde{R}^{(0)}, \widetilde{R}^{(1)}, \ldots) \in \widetilde{\mathcal{T}} : \widetilde{R}_{IJKL,M_1\cdots M_r} = \widetilde{R}_{IJKL,(M_1\cdots M_r)}\}$$

$$T\widetilde{\mathcal{R}}^N = \{(\widetilde{R}_{IJKL,M_1\cdots M_r}) \in \widetilde{\mathcal{T}}^N : \widetilde{R}_{IJKL,M_1\cdots M_r} = \widetilde{R}_{IJKL,(M_1\cdots M_r)}\}$$

$$\widetilde{\sigma}^N = \{(\widetilde{R}_{IJKL,M_1\cdots M_r}) \in \widetilde{\tau}^N : \widetilde{R}_{IJKL,M_1\cdots M_r} = \widetilde{R}_{IJKL,(M_1\cdots M_r)}\}.$$

Then $T\widetilde{\mathcal{R}}^N = \bigoplus_{M=0}^N \widetilde{\sigma}^M$ and $T\widetilde{\mathcal{R}} = \prod_{M=0}^\infty \widetilde{\sigma}^M$. Note that in the presence of the condition $\widetilde{R}_{IJKL,M_1\cdots M_r} = \widetilde{R}_{IJKL,(M_1\cdots M_r)}$, condition (6) of Definition 8.8 becomes

$$\widetilde{R}_{IJKL,M_1\cdots M_r 0} = -(r+2)\widetilde{R}_{IJKL,M_1\cdots M_r} \tag{8.4}$$

and is a consequence of (2) and (5). Note also that if $(\widetilde{R}_{IJKL,M_1\cdots M_r}) \in \widetilde{\sigma}^N$, then $\widetilde{h}^{LM_s}\widetilde{R}_{IJKL,M_1\cdots M_r} = 0$ and $\widetilde{h}^{M_sM_t}\widetilde{R}_{IJKL,M_1\cdots M_r} = 0$.

For $w \in \mathbb{C}$, let $\sigma_w : P \to \mathbb{C}$ denote the character $\sigma_w(p) = a^{-w}$. Since $P \subset O(\widetilde{h})$ and P preserves e_0 up to scale, it is easily seen that $\widetilde{\mathcal{T}}$, $\widetilde{\mathcal{R}}$ and $T\widetilde{\mathcal{R}}$ are invariant subsets of the P-space $\prod_{r=0}^{\infty}\left(\bigotimes^{4+r}\mathbb{R}^{n+2*}\otimes\sigma_{-2-r}\right)$, where $\mathbb{R}^{n+2}$ denotes the standard representation of $P \subset GL(n+2,\mathbb{R})$. These inclusions therefore define actions of P on these spaces. These actions of P do not preserve strength, but because P consists of block upper-triangular matrices, a component of $p.(\widetilde{R})$ depends only on components of $\widetilde{R}$ of no greater strength. So for $N \geq 0$ there are also actions of P on $\widetilde{\mathcal{T}}^N$, $\widetilde{\mathcal{R}}^N$ and $T\widetilde{\mathcal{R}}^N$. An easy computation shows that the element $p_a \in P$ acts by multiplying a component of strength S by a^{S-2}.

We next define our main object of interest. If g is a metric defined in a neighborhood of $0 \in \mathbb{R}^n$, we construct a straight ambient metric in normal form for g as in Chapter 3. We then evaluate the covariant derivatives of curvature of the ambient metric at $\rho = 0, t = 1$ as described in Chapter 6. If n is odd, the values of all components of these covariant derivatives at the origin depend only on the derivatives of g at the origin, while if n is even, this is true for components of strength at most $n+1$ by Proposition 6.2. If $g_{ij}(0) = h_{ij}$, then $\widetilde{g}_{IJ} = \widetilde{h}_{IJ}$ at $\rho = 0$, $t = 1$, $x = 0$. In this case, Propositions 6.1 and 6.4 show that the resulting lists of components satisfy the relations of Definition 8.8. This procedure therefore defines a map $c : \mathcal{N}_c \to \widetilde{\mathcal{R}}$ for n odd, and $c : \mathcal{N}_c \to \widetilde{\mathcal{R}}^{n-3}$ for n even. Since the conformal curvature tensors are natural polynomial invariants of the metric g, c is a polynomial map.

Proposition 8.9. *The map $c : \mathcal{N}_c \to \widetilde{\mathcal{R}}$ (or $\widetilde{\mathcal{R}}^{n-3}$ if n is even) is equivariant with respect to the P-actions.*

Proof. Recall that the action of P on $\mathcal{N}_c$ is given by $p.g = (\varphi^{-1})^*(\Omega^2 g)$, where φ and Ω are determined to map g back to conformal normal form given the initial normalizations defined by p; see the discussion preceding Proposition 8.7 above. By naturality of the conformal curvature tensors, $c((\varphi^{-1})^*(\Omega^2 g)) = (\varphi'(0)^{-1})^*(c(\Omega^2 g))$, where $(\varphi'(0)^{-1})^*$ on the right-hand side is interpreted as the pullback in the indices between 1 and n of each of the tensors in the list, leaving the 0 and ∞ indices alone. And $c(\Omega^2 g)$ is given by Proposition 6.5. We use these observations to check for each of the generating subgroups $O(h)$, $\mathbb{R}_+$ and $\mathbb{R}^n$ of P that $c(p.g) = p.(c(g))$, where the P-action on the right-hand side is that on $\prod_{r=0}^{\infty}\left(\bigotimes^{4+r}\mathbb{R}^{n+2*}\otimes\sigma_{-2-r}\right)$.

For $p = p_m$, we have $\Omega = 1$ and $\varphi(x) = mx$, so $c(p_m.g) = (m^{-1})^*c(g)$ is obtained from $c(g)$ by transforming covariantly under $O(h)$ the indices between 1 and n. But this is precisely how p_m acts on $\prod_{r=0}^{\infty}\bigotimes^{4+r}\mathbb{R}^{n+2*}$.

For $p = p_a$, we have $\varphi(x) = a^{-1}x$ and $\Omega = a^{-2}$. By Proposition 6.5, the component $\widetilde{R}_{IJKL,M_1\cdots M_r}$ for $a^{-2}g$ is that for g multiplied by $a^{2s_\infty-2}$. Since $(\varphi'(0)^{-1})^*$ acts by multiplying this component by a^{s_M}, it follows that the components of $c(p_a.g)$ are those of $c(g)$ multiplied by $a^{s_M+2s_\infty-2} = a^{S-2}$, where $S = \|IJKLM_1\cdots M_r\|$. But we noted above that this is precisely how p_a acts on $\prod_{r=0}^{\infty}\left(\bigotimes^{4+r}\mathbb{R}^{n+2*}\otimes\sigma_{-2-r}\right)$.

Finally, for $p = p_b$, we have $\varphi'(0) = I$ and $\Omega = 1 - b_i x^i + O(|x|^2)$, so that the components of $c((\varphi^{-1})^*(\Omega^2 g))$ are given by Proposition 6.5 with $\Upsilon_i = -b_i$. But this is precisely how p_b acts on $\prod_{r=0}^{\infty}\bigotimes^{4+r}\mathbb{R}^{n+2*}$. □

Let us examine more carefully the equivariance of c with respect to the subgroup $\mathbb{R}_+ \subset P$. The component $\widetilde{R}_{IJKL,M_1\cdots M_r}$ of $c(g)$ is a polynomial in the components of the $g^{(s)}$, and this equivariance says that when the $g^{(s)}$ are replaced by $a^s g^{(s)}$, then $\widetilde{R}_{IJKL,M_1\cdots M_r}$ is multiplied by a^{S-2} with $S = \|IJKLM_1\cdots M_r\|$. In particular, $\widetilde{R}_{IJKL,M_1\cdots M_r}$ can only involve $g^{(s)}$ for $s \le S-2$. This implies that for each $N \ge 0$ (satisying also $N \le n-3$ for n even), c induces a P-equivariant polynomial map $c^N : \mathcal{N}_c^{N+2} \to \widetilde{\mathcal{R}}^N$. Clearly these induced maps satisfy the compatibility conditions $c^{N-1}\pi_{N+2} = \widetilde{\pi}_N c^N$, where $\pi_N : \mathcal{N}_c^N \to \mathcal{N}_c^{N-1}$ and $\widetilde{\pi}_N : \widetilde{\mathcal{R}}^N \to \widetilde{\mathcal{R}}^{N-1}$ are the natural projections.

The main result of this chapter is the following jet isomorphism theorem.

Theorem 8.10. *Let $N \ge 0$ and assume that $N \le n-3$ if n is even. Then $\widetilde{\mathcal{R}}^N$ is a smooth submanifold of $\widetilde{\mathcal{T}}^N$ whose tangent space at 0 is $T\widetilde{\mathcal{R}}^N$, and the map $c^N : \mathcal{N}_c^{N+2} \to \widetilde{\mathcal{R}}^N$ is a P-equivariant polynomial equivalence.*

For n odd, it follows that $c : \mathcal{N}_c \to \widetilde{\mathcal{R}}$ is a P-equivariant bijection since c is the projective limit of the c^N.

It will be convenient in the proof of Theorem 8.10 to use Theorem 8.3 to realize $\mathcal{N}_c$ in terms of curvature tensors on $\mathbb{R}^n$ rather than Taylor coefficients of metrics. So we make the following definition. Recall the space $\mathcal{R}$ introduced in Definition 8.2.

Definition 8.11. Define the space $\mathcal{R}_c \subset \mathcal{R}$ to be the subset consisting of lists of tensors $(R^{(0)}, R^{(1)}, \ldots)$ satisfying in addition to the conditions in Definition 8.2 the following: for each $r \ge 0$,

$$\mathrm{Sym}(h^{ik}R^{(r)}_{ijkl,m_1\cdots m_r}) = 0. \tag{8.5}$$

Here Sym refers to the symmetrization over the free indices $jlm_1\cdots m_r$. For $N \ge 0$, by $\mathcal{R}_c^N$ we will denote the corresponding set of truncated lists $(R^{(0)}, R^{(1)}, \ldots, R^{(N)})$.

For $r, N \ge 0$, we define also the following finite-dimensional vector spaces:

$$\begin{aligned}
\tau^r &= \{R^{(r)} \in \bigwedge^2 \mathbb{R}^{n*} \otimes \bigwedge^2 \mathbb{R}^{n*} \otimes \bigotimes^r \mathbb{R}^{n*} : \\
&\qquad\qquad\qquad (8.5) \text{ and } (1), (2) \text{ of Definition 8.2 hold}\} \\
\sigma^r &= \{R^{(r)} \in \tau^r : R_{ijkl,m_1\cdots m_r} = R_{ijkl,(m_1\cdots m_r)}\} \\
\mathcal{T}^N &= \textstyle\bigoplus_{r=0}^N \tau^r \\
T\mathcal{R}_c^N &= \textstyle\bigoplus_{r=0}^N \sigma^r \subset \mathcal{T}^N.
\end{aligned}$$

The bijection $\mathcal{N} \to \mathcal{R}$ asserted by Theorem 8.3 clearly restricts to a bijection $\mathcal{N}_c \to \mathcal{R}_c$ whose truncated maps $\mathcal{N}_c^{N+2} \to \mathcal{R}_c^N$ are polynomial equivalences. By composition, we can regard c and the c^N as defined on the corresponding $\mathcal{R}_c$ and $\mathcal{R}_c^N$. In the following, we will not have occasion to refer to $\mathcal{N}_c$ and $\mathcal{N}_c^{N+2}$, so no confusion should arise from henceforth using the same symbols c and c^N for the maps defined on $\mathcal{R}_c$ and $\mathcal{R}_c^N$. We can transfer the action of P on $\mathcal{N}_c$ to $\mathcal{R}_c$. The element p_a acts on $\mathcal{R}_c$ by multiplying $R^{(r)}$ by a^{r+2}; this same prescription gives an $\mathbb{R}_+$ action on $\mathcal{T}^N$ for $N \geq 0$. The $\mathbb{R}_+$-equivariance of c^N implies that a component $\widetilde{R}_{IJKL,M_1\cdots M_r}$ of $c^N(R)$ with $\|IJKLM_1\cdots M_r\| = N+4$ can be written as a linear combination of components of $R^{(N)}$ plus quadratic and higher terms in the components of the $R^{(r)}$ with $r \leq N-2$.

Our starting point for the proof of Theorem 8.10 is the following lemma.

Lemma 8.12. *For each $N \geq 0$, the subset $\mathcal{R}_c^N \subset \mathcal{T}^N$ is a smooth submanifold whose tangent space at 0 is $T\mathcal{R}_c^N$.*

Proof. We will show that for each $N \geq 1$, there is a polynomial equivalence $\Phi^N : \mathcal{T}^N \to \mathcal{T}^N$ satisfying $d\Phi^N(0) = I$ and $\Phi^N(\mathcal{R}_c^{N-1} \times \sigma^N) = \mathcal{R}_c^N$. Upon iterating this statement and using $\sigma^0 = \mathcal{R}_c^0 = T\mathcal{R}_c^0$, we conclude the existence of a polynomial equivalence : $\mathcal{T}^N \to \mathcal{T}^N$ whose differential at 0 is the identity, and which maps $T\mathcal{R}_c^N \to \mathcal{R}_c^N$. The desired conclusion follows immediately.

When reformulated in terms of the spaces $\mathcal{R}_c^N$, Lemma 8.6 asserts the existence for each $N \geq 1$ of a polynomial map $\eta^N : \mathcal{T}^{N-1} \to \bigotimes^{N+4} \mathbb{R}^{n*}$ with zero constant and linear term, such that the map $\Lambda^N : R \to (R, \eta^N(R))$ sends $\mathcal{R}_c^{N-1} \to \mathcal{R}_c^N$. Here R denotes the list constituting an element of $\mathcal{T}^{N-1}$. There is no loss of generality in assuming that $\eta^N(\mathcal{T}^{N-1}) \subset \tau^N$ so that $\Lambda^N : \mathcal{T}^{N-1} \to \mathcal{T}^N$. Define $\Phi^N : \mathcal{T}^N \to \mathcal{T}^N$ by $\Phi^N(R, R^{(N)}) = (R, R^{(N)} + \eta^N(R))$. It is evident from the form of the relations defining $\mathcal{R}_c^N$ that $\Phi^N(\mathcal{R}_c^{N-1} \times \sigma^N) = \mathcal{R}_c^N$, and clearly $(\Phi^N)^{-1}(R, R^{(N)}) = (R, R^{(N)} - \eta^N(R))$. □

We remark that the same proof could have been carried out in terms of the spaces of normal form coefficients, and shows that the subset $\mathcal{N}_c^N \subset \mathcal{N}^N$ is a smooth submanifold whose tangent space at 0 is obtained by linearizing the equation obtained by writing (8.5) in terms of the normal form coefficients $g_{ij,k_1\cdots k_r}$.

At this point we do not know that $\widetilde{\mathcal{R}}^N$ is a submanifold of $\widetilde{\mathcal{T}}^N$ with tangent space $T\widetilde{\mathcal{R}}^N$, but it is clear that the tangent vector at 0 to a smooth curve in $\widetilde{\mathcal{R}}^N$ must lie in $T\widetilde{\mathcal{R}}^N$. So we conclude for the differential of c^N at the origin that $dc^N : T\mathcal{R}_c^N \to T\widetilde{\mathcal{R}}^N$. The differentiation of the action of P on $\mathcal{R}_c^N$ gives a linear action of P on $T\mathcal{R}_c^N$, and $dc^N : T\mathcal{R}_c^N \to T\widetilde{\mathcal{R}}^N$ is P-equivariant. By $\mathbb{R}_+$-equivariance and linearity of dc^N, it follows that dc^N decomposes as a direct sum of maps $\sigma^M \to \widetilde{\sigma}^M$ for $0 \leq M \leq N$. By the compatibility of the c^N as N varies, the map $\sigma^M \to \widetilde{\sigma}^M$ is independent of the choice of $N \geq M$, so we may as well denote it as $dc^N : \sigma^N \to \widetilde{\sigma}^N$. The main algebraic fact on which rests the proof of Theorem 8.10 is the following.

Proposition 8.13. *For $N \geq 0$ (and $N \leq n-3$ if n is even), $dc^N : \sigma^N \to \widetilde{\sigma}^N$ is an isomorphism.*

Proof of Theorem 8.10 using Proposition 8.13. We prove by induction on N that there is a polynomial equivalence $\widetilde{\Phi}^N : \widetilde{\mathcal{T}}^N \to \widetilde{\mathcal{T}}^N$ satisfying $d\widetilde{\Phi}^N(0) = I$ and $\widetilde{\Phi}^N(\widetilde{\mathcal{R}}^{N-1} \times \widetilde{\sigma}^N) = \widetilde{\mathcal{R}}^N$, and that $c^N : \mathcal{R}_c^N \to \widetilde{\mathcal{R}}^N$ is a polynomial equivalence. Just as in the proof of Lemma 8.12, iterating the first statement provides a polynomial equivalence : $\widetilde{\mathcal{T}}^N \to \widetilde{\mathcal{T}}^N$ whose differential at 0 is the identity and which maps $T\widetilde{\mathcal{R}}^N \to \widetilde{\mathcal{R}}^N$, from which follows the first statement of Theorem 8.10.

For $N = 0$, we can take $\widetilde{\Phi}^N$ to be the identity. Since $\mathcal{R}_c^0 = T\mathcal{R}_c^0 = \sigma^0$, $\widetilde{\mathcal{R}}^0 = T\widetilde{\mathcal{R}}^0 = \widetilde{\sigma}^0$, and c^0 is linear and can be identified with dc^0, the second statement is immediate from Proposition 8.13.

Suppose for some $N \geq 1$ that we have the polynomial equivalence $\widetilde{\Phi}^{N-1}$ and we know that c^{N-1} is a polynomial equivalence. Recall the polynomial maps $\eta^N : \mathcal{T}^{N-1} \to \tau^N$, $\Lambda^N : \mathcal{T}^{N-1} \to \mathcal{T}^N$ and $\Phi^N : \mathcal{T}^N \to \mathcal{T}^N$ constructed in Lemma 8.6 and Lemma 8.12. By the induction hypothesis that c^{N-1} is a polynomial equivalence, we conclude that there is a polynomial map $\widetilde{\Lambda}^N : \widetilde{\mathcal{T}}^{N-1} \to \widetilde{\mathcal{T}}^N$ such that $\widetilde{\Lambda}^N(\widetilde{\mathcal{R}}^{N-1}) \subset \widetilde{\mathcal{R}}^N$ and such that the diagram

$$\begin{array}{ccc} \mathcal{R}_c^{N-1} & \xrightarrow{\Lambda^N} & \mathcal{R}_c^N \\ \Big\downarrow {\scriptstyle c^{N-1}} & & \Big\downarrow {\scriptstyle c^N} \\ \widetilde{\mathcal{R}}^{N-1} & \xrightarrow{\widetilde{\Lambda}^N} & \widetilde{\mathcal{R}}^N \end{array}$$

commutes. Using the compatibility of c^{N-1} and c^N and the form of the map Λ^N, one sees that $\widetilde{\Lambda}^N$ can be taken to have the form $\widetilde{\Lambda}^N(\widetilde{R}) = (\widetilde{R}, \widetilde{\eta}^N(\widetilde{R}))$ where $\widetilde{\eta}^N : \widetilde{\mathcal{T}}^{N-1} \to \widetilde{\tau}^N$ has no constant or linear terms. Now define the map $\widetilde{\Phi}^N : \widetilde{\mathcal{T}}^N \to \widetilde{\mathcal{T}}^N$ by $\widetilde{\Phi}^N(\widetilde{R}, \widetilde{R}^{(N)}) = (\widetilde{R}, \widetilde{R}^{(N)} + \widetilde{\eta}^N(\widetilde{R}))$. Clearly $d\widetilde{\Phi}^N(0) = I$, and $\widetilde{\Phi}^N(\widetilde{\mathcal{R}}^{N-1} \times \widetilde{\sigma}^N) = \widetilde{\mathcal{R}}^N$ by the form of the relations defining $\widetilde{\mathcal{R}}^N$. It is a straightforward matter to check that the diagram

$$\begin{array}{ccc} \mathcal{R}_c^{N-1} \times \sigma^N & \xrightarrow{\Phi^N} & \mathcal{R}_c^N \\ \big\downarrow {\scriptstyle c^{N-1}\times dc^N} & & \big\downarrow {\scriptstyle c^N} \\ \widetilde{\mathcal{R}}^{N-1} \times \widetilde{\sigma}^N & \xrightarrow{\widetilde{\Phi}^N} & \widetilde{\mathcal{R}}^N \end{array}$$

commutes. By the induction hypothesis and Proposition 8.13, the vertical map on the left is a polynomial equivalence. We conclude that c^N is also a polynomial equivalence, completing the induction step. □

Proof of Proposition 8.13. The proof has two parts. We will first construct an injective map $L : \widetilde{\sigma}^N \to \sigma^N$. We will then show that $dc^N : \sigma^N \to \widetilde{\sigma}^N$ is injective. These statements together imply that $\dim(\sigma^N) = \dim(\widetilde{\sigma}^N)$, from which it then follows that dc^N is an isomorphsim.

Let $(\widetilde{R}_{IJKL,M_1\cdots M_r}) \in \widetilde{\sigma}^N$. We can consider the components $\widetilde{R}_{ijkl,m_1\cdots m_N}$ in which all the indices lie between 1 and n. This defines a map $L : \widetilde{\sigma}^N \to \bigwedge^2 \mathbb{R}^{n*} \otimes \bigwedge^2 \mathbb{R}^{n*} \otimes \bigodot^N \mathbb{R}^{n*}$, and clearly everything in the range of L satisfies conditions (1) and (2) of Definition 8.2. We claim that everything in the range of L also satisfies (8.5), so that $L : \widetilde{\sigma}^N \to \sigma^N$. Condition (3) of Definition 8.8 implies $\widetilde{h}^{IK}\widetilde{R}_{IjKl,m_1\cdots m_N} = 0$. This can be written as

$$\widetilde{R}_{j\infty l0,m_1\cdots m_N} + \widetilde{R}_{l\infty j0,m_1\cdots m_N} + h^{ik}\widetilde{R}_{ijkl,m_1\cdots m_N} = 0.$$

If we apply condition (5) of Definition 8.8 to $\widetilde{R}_{j\infty l0,m_1\cdots m_N}$ and then symmetrize over $jlm_1\cdots m_N$, the result is 0 by the skew symmetry of $\widetilde{R}^{(N)}$ in the second pair of indices. Similarly for $\widetilde{R}_{l\infty j0,m_1\cdots m_N}$. It follows that the symmetrization of $h^{ik}\widetilde{R}_{ijkl,m_1\cdots m_N}$ vanishes. This proves that (8.5) holds.

Next we show that L is injective. We claim that for $(\widetilde{R}_{IJKL,M_1\cdots M_r}) \in \widetilde{\sigma}^N$, any component $\widetilde{R}_{IJKL,M_1\cdots M_r}$ can be written as a linear combination of components in which none of the indices $IJKLM_1\cdots M_r$ is ∞. We first show that any component can be written as a linear combination of components in which none of $IJKL$ is ∞. To see this, note that (8.4) and $\widetilde{h}^{AB}\widetilde{R}_{IJKA,BM_1\cdots M_r} = 0$ imply

$$\begin{aligned} -(r+2)\widetilde{R}_{IJK\infty,M_1\cdots M_r} &= \widetilde{R}_{IJK\infty,0M_1\cdots M_r} \\ &= -\widetilde{R}_{IJK0,\infty M_1\cdots M_r} - h^{ab}\widetilde{R}_{IJKa,bM_1\cdots M_r}. \end{aligned}$$

Thus a component $\widetilde{R}_{IJKL,M_1\cdots M_r}$ in which $L = \infty$ can be rewritten as a linear combination of components in which $L \neq \infty$ and IJK remain unchanged. Repeating this procedure allows the removal of any ∞'s in $IJKL$.

The same method allows the removal of ∞'s in $M_1 \cdots M_r$: from (8.4) and $\widetilde{h}^{AB}\widetilde{R}_{IJKL,ABM_2\cdots M_r} = 0$, one has

$$-2(r+2)\widetilde{R}_{IJKL,\infty M_2\cdots M_r} = 2\widetilde{R}_{IJKL,0\infty M_2\cdots M_r} = -h^{ab}\widetilde{R}_{IJKL,abM_2\cdots M_r}.$$

Thus all ∞'s can be removed as indices. Now (8.4) and (5) of Definition 8.8 can be used to remove any 0's as indices at the expense of permuting the remaining indices between 1 and n. It follows that any component $\widetilde{R}_{IJKL,M_1\cdots M_r}$ can be written as a linear combination of components in which all indices are between 1 and n. Thus L is injective.

It remains to prove that dc^N is injective. If $R_{ijkl,m_1\cdots m_N} \in \sigma^N$, we set $R_{jl,m_1\cdots m_N} = h^{ik}R_{ijkl,m_1\cdots m_N}$, $R_{m_1\cdots m_N} = h^{jl}R_{jl,m_1\cdots m_N}$ and

$$(n-2)P_{jk,m_1\cdots m_N} = R_{jk,m_1\cdots m_N} - R_{m_1\cdots m_N}h_{jk}/2(n-1).$$

We also denote by $W_{ijkl,m_1\cdots m_N} \in \bigwedge^2\mathbb{R}^{n*} \otimes \bigwedge^2\mathbb{R}^{n*} \otimes \bigodot^N\mathbb{R}^{n*}$ the tensor obtained by taking the trace-free part in $ijkl$ while ignoring $m_1\cdots m_N$:

$$W_{ijkl,m_1\cdots m_N} = R_{ijkl,m_1\cdots m_N} - 2h_{l[j}P_{i]k,m_1\cdots m_N}h_{jl} + 2h_{k[j}P_{i]l,m_1\cdots m_N}. \quad (8.6)$$

Then $W_{ijkl,m_1\cdots m_N}$ satisfies (1) of Definition 8.2 but not necessarily (2). We also define

$$C_{jkl,m_1\cdots m_{N-1}} = 2P_{j[k,l]m_1\cdots m_{N-1}}.$$

Contracting the second Bianchi identity (2) of Definition 8.2 in the usual way shows that $W^i{}_{jkl,im_1\cdots m_{N-1}} = (3-n)C_{jkl,m_1\cdots m_{N-1}}$ and $P^i{}_{j,im_1\cdots m_{N-1}} = P^i{}_{i,jm_1\cdots m_{N-1}}$.

Lemma 8.14. *Let $R_{ijkl,m_1\cdots m_N} \in \sigma^N$. If $n \geq 4$ and $W_{ijkl,m_1\cdots m_N} = 0$, then $R_{ijkl,m_1\cdots m_N} = 0$. If $n = 3$ and $C_{jkl,m_1\cdots m_{N-1}} = 0$, then $R_{ijkl,m_1\cdots m_N} = 0$.*

Proof. If $N = 0$, this follows from the decomposition of the curvature tensor into its Weyl piece and its Ricci piece. Suppose $N \geq 1$. If $n \geq 4$, the contracted Bianchi identity above shows that $C_{jkl,m_1\cdots m_{N-1}} = 0$, which is our hypothesis if $n = 3$. (The hypothesis $W_{ijkl,m_1\cdots m_N} = 0$ for $n \geq 4$ is of course automatic for $n = 3$.) Thus we conclude for any n that $P_{j[k,l]m_1\cdots m_{N-1}} = 0$, so $P_{ij,m_1\cdots m_N} = P_{(ij,m_1\cdots m_N)}$. Since $R_{ijkl,m_1\cdots m_N} \in \sigma^N$, we also have $R_{(ij,m_1\cdots m_N)} = 0$. Therefore

$$(n-2)P_{ij,m_1\cdots m_N} = (n-2)P_{(ij,m_1\cdots m_N)} = -R_{(m_1\cdots m_N}h_{ij)}/2(n-1).$$

Now $h^{ij}P_{(ij,m_1\cdots m_N)} = h^{ij}P_{ij,m_1\cdots m_N} = R_{m_1\cdots m_N}/2(n-1)$. Hence the symmetric tensor $P = P_{(ij,m_1\cdots m_N)}$ is in the kernel of the operator $P \to (n-2)P + \mathrm{Sym}(\mathrm{tr}(P)h)$. This operator is injective on symmetric tensors, so we conclude that $P_{ij,m_1\cdots m_N} = 0$. The conclusion now follows from (8.6). □

We will prove that dc^N is injective by showing that if $R_{ijkl,m_1\cdots m_N} \in \ker(dc^N)$, then the hypotheses of Lemma 8.14 hold. To get the flavor of the argument, consider first the cases $N = 0, 1$. Now σ^0 is the space of trace-free curvature tensors. In Chapter 6, we found that $\widetilde{R}_{ijkl} = W_{ijkl}$ is the Weyl piece of such a curvature tensor. So c^0 is linear and is obviously injective. In Chapter 6, we also calculated the curvature components $\widetilde{R}_{\infty jkl} = C_{jkl}$ and $\widetilde{R}_{ijkl,m}$ (see (6.3)). We see that c^1 is also linear, so can be identified with dc^1. If $R_{ijkl,m} \in \ker(dc^1)$, we first conclude by considering $\widetilde{R}_{\infty jkl}$ that $C_{jkl} = 0$, and then by considering $\widetilde{R}_{ijkl,m}$ that $W_{ijkl,m} = 0$ (for $n \geq 4$), as desired.

For the general case we need to understand the relation between covariant derivatives with respect to the ambient metric and covariant derivatives with respect to a representative g on M. Recall that the conformal curvature tensors are tensors on M defined by evaluating components of $\widetilde{R}_{IJKL,M_1\cdots M_r}$ at $\rho = 0$ and $t = 1$. We can take further covariant derivatives of such a tensor with respect to g. We will denote by $\widetilde{R}_{IJKL,M_1\cdots M_r|p_1\cdots p_s}$ the tensor on M obtained by such further covariant differentiations. For example, $\widetilde{R}_{ijkl,|m} = W_{ijkl,m}$, whereas $\widetilde{R}_{ijkl,m}$ is the tensor V_{ijklm} given by (6.3). An inspection of (3.16) shows the relation between $\widetilde{R}_{IJKL,M_1\cdots M_r p}$ and $\widetilde{R}_{IJKL,M_1\cdots M_r|p}$. Recalling that $g'_{ij} = 2P_{ij}$ at $\rho = 0$, one sees that $\widetilde{R}_{IJKL,M_1\cdots M_r p} - \widetilde{R}_{IJKL,M_1\cdots M_r|p}$ is a linear combination of components of $\widetilde{\nabla}^r\widetilde{R}$, possibly multiplied by components of g, plus quadratic terms in curvature. Iterating, it follows that $\widetilde{R}_{IJKL,M_1\cdots M_r p_1\cdots p_s} - \widetilde{R}_{IJKL,M_1\cdots M_r|p_1\cdots p_s}$ is a linear combination of terms of the form $\widetilde{R}_{ABCD,F_1\cdots F_u|q_1\cdots q_t}$ with $t < s$, possibly multiplied by components of g, plus nonlinear terms in curvature.

Consider now the map $dc^N : \sigma^N \to \widetilde{\sigma}^N$. The symbol $\widetilde{R}_{IJKL,M_1\cdots M_r}$ with $\|IJKLM_1\cdots M_r\| = N + 4$ is now to be interpreted as the linear function of the $R_{ijkl,m_1\cdots m_N}$ obtained by applying dc^N. Similarly, we now interpret the symbol $\widetilde{R}_{IJKL,M_1\cdots M_r|p_1\cdots p_s}$ for $\|IJKLM_1\cdots M_r\| + s = N + 4$ as a linear function of the $R_{ijkl,m_1\cdots m_N}$. Suppose that $R_{ijkl,m_1\cdots m_N} \in \ker(dc^N)$. We claim that $\widetilde{R}_{IJKL,M_1\cdots M_r|p_1\cdots p_s} = 0$ for $\|IJKLM_1\cdots M_r\| + s = N + 4$. The proof is by induction on s. For $s = 0$, this is just the hypothesis that $R_{ijkl,m_1\cdots m_N} \in \ker(dc^N)$. Since we are considering the linearization dc^N, the quadratic terms may be ignored in the relation derived in the previous paragraph between ambient covariant derivatives and covariant derivatives on M. A moment's thought shows that this relation provides the induction step to increase s by 1.

Taking $r = 0$, we conclude that $\widetilde{R}_{IJKL,|m_1\cdots m_s} = 0$ for $\|IJKL\| + s = N + 4$. For $IJKL = ijkl$ we obtain $W_{ijkl,m_1\cdots m_N} = 0$ and for $IJKL = \infty jkl$ we obtain $C_{jkl,m_1\cdots m_{N-1}} = 0$. Lemma 8.14 then shows that $R_{ijkl,m_1\cdots m_N} = 0$ as desired. □

Chapter Nine

Scalar Invariants

The jet isomorphism theorem 8.10 reduces the study of local invariants of conformal structures to the study of P-invariants of $\widetilde{\mathcal{R}}$ (we must of course impose the usual finite-order truncation for n even). An invariant theory for scalar P-invariants of $T\widetilde{\mathcal{R}}$ was developed in [BEGr]. In this chapter we show how to derive a characterization of scalar invariants of conformal structures by reduction to the relevant results of [BEGr].

Recall that a scalar invariant $I(g)$ of metrics of signature (p,q) is a polynomial in the variables $(\partial^\alpha g_{ij})_{|\alpha|\geq 0}$ and $|\det g_{ij}|^{-1/2}$, which is coordinate-free in the sense that its value is independent of orientation-preserving changes of the coordinates used to express and differentiate g. Such a scalar invariant of metrics is said to be *even* if it is also unchanged under orientation-reversing changes of coordinates, and *odd* if it changes sign under orientation-reversing coordinate changes.

It follows from the jet isomorphism theorem for pseudo-Riemannian geometry and Weyl's classical invariant theory that every scalar invariant of metrics is a linear combination of complete contractions

$$\begin{gathered} \operatorname{contr}(\nabla^{r_1} R \otimes \cdots \otimes \nabla^{r_L} R) \\ \operatorname{contr}(\mu \otimes \nabla^{r_1} R \otimes \cdots \otimes \nabla^{r_L} R), \end{gathered} \tag{9.1}$$

where $\mu_{i_1\cdots i_n} = |\det g|^{1/2}\varepsilon_{i_1\cdots i_n}$ is the volume form with respect to a chosen orientation and the contractions are with respect to g. Here $\varepsilon_{i_1\cdots i_n}$ denotes the sign of the permutation. Complete contractions of the first type are even and the second type are odd. Such an invariant of metrics is said to be conformally invariant of weight w if $I(\Omega^2 g) = \Omega^w I(g)$ for smooth positive functions Ω. Under a constant rescaling $g \to a^2 g$, the complete contractions above are multiplied by $a^{-2L-\sum r_i}$. Since the total number of contracted indices must be even, it follows that the weight of a nonzero even scalar conformal invariant must be a negative even integer, and that of a nonzero odd scalar conformal invariant must be a negative integer which agrees with $n \mod 2$.

Scalar conformal invariants can be constructed quite simply via the ambient metric. Consider complete contractions

$$\begin{aligned} &\operatorname{contr}(\widetilde{\nabla}^{r_1}\widetilde{R}\otimes\cdots\otimes\widetilde{\nabla}^{r_L}\widetilde{R}) \\ &\operatorname{contr}(\widetilde{\mu}\otimes\widetilde{\nabla}^{r_1}\widetilde{R}\otimes\cdots\otimes\widetilde{\nabla}^{r_L}\widetilde{R}) \\ &\operatorname{contr}(\widetilde{\mu}_0\otimes\widetilde{\nabla}^{r_1}\widetilde{R}\otimes\cdots\otimes\widetilde{\nabla}^{r_L}\widetilde{R}), \end{aligned} \tag{9.2}$$

where now $\widetilde{R}$ is the curvature tensor of a straight ambient metric $\widetilde{g}$ for $[g]$, $\widetilde{\mu}$ denotes the volume form of $\widetilde{g}$ with respect to an orientation on $\widetilde{\mathcal{G}}$ induced from a choice of orientation on M, $\widetilde{\mu}_0 = T \lrcorner \widetilde{\mu}$, and the contractions are taken with respect to $\widetilde{g}$. When evaluated for a specific ambient metric, these contractions define functions on $\widetilde{\mathcal{G}}$. The homogeneity of $\widetilde{g}$ and T imply that the functions defined by contractions of the first two types are homogeneous of degree $-2L - \sum r_i$ with respect to the dilations δ_s, while those of the third type are homogeneous of degree $1 - 2L - \sum r_i$. Their restrictions to $\mathcal{G} \subset \widetilde{\mathcal{G}}$ are independent of the diffeomorphism ambiguity of $\widetilde{g}$ (we restrict to orientation-preserving diffeomorphisms for the second and third types) since the diffeomorphism restricts to the identity on $\mathcal{G}$. If n is odd, then the restriction of any such contraction to $\mathcal{G}$ is also clearly independent of the infinite-order ambiguity in the ambient metric, so depends only on $[g]$. A representative metric g defines a section of $\mathcal{G}$, so composing with this section gives a function $I(g)$ on M. The homogeneity of the contraction as a function on $\mathcal{G}$ implies that $I(\Omega^2 g) = \Omega^w I(g)$, where $-w = 2L + \sum r_i$ for the first two and $-w = 2L + \sum r_i - 1$ for the third. If we take $\widetilde{g}$ to be in normal form relative to g and recall the discussion of conformal curvature tensors in Chapter 6, it follows that in local coordinates $I(g)$ does indeed have the required polynomial dependence on the Taylor coefficients of g. Thus $I(g)$ is a scalar conformal invariant of weight w. This proves the following proposition in case n is odd.

Proposition 9.1. *If n is odd, the complete contractions* (9.2) *define scalar conformal invariants. The first is even and has $-w = 2L + \sum r_i$; the second and third are odd and have $-w = 2L + \sum r_i$ and $-w = 2L + \sum r_i - 1$. If n is even, the same statements are true with the restrictions $L \geq 2$ and $-w \leq n + 2$ for the first contraction, and $-w \leq 2n - 2$ for the last two.*

By a Weyl conformal invariant of metrics, we will mean a linear combination of complete contractions (9.2), all of which have the same weight, and which satisfy the restrictions of Proposition 9.1 if n is even. Every Weyl invariant can be written as a sum of an even Weyl invariant and an odd Weyl invariant.

Before giving the proof of Proposition 9.1 for n even, we make some observations concerning odd invariants. The Bianchi and Ricci identities imply that the skew-symmetrization over any three indices of $\nabla^r R$ can be written as a quadratic expression in the $\nabla^{r'} R$ with $r' \leq r - 2$. An odd contraction in (9.1) with $L < n/2$ necessarily has at least three of the indices of at least one of the $\nabla^{r_i} R$ contracted against indices in μ. So by induction, it follows that an odd contraction in (9.1) can be written as a linear combination of contractions of the same form but with $L \geq n/2$. Similarly, contractions of the second and third types in (9.2) can be written as linear combinations of contractions of the same type and the same weight, but with $L \geq (n+2)/2$ for the second type and $L \geq (n+1)/2$ for the third type. This leads to the following theorem.

Theorem 9.2. *There are no nonzero odd scalar conformal invariants with $-w < n$. There are no nonzero odd Weyl invariants with $-w < n+1$.*

Proof. The first statement is clear from the above observations since $-w = 2L + \sum r_i$. The same reasoning shows that there are no nonzero contractions of the second type in (9.2) with $-w < n+2$. For contractions of the third type, one has $-w = 2L + \sum r_i - 1$, which is $< n+1$ only if $L = (n+1)/2$ (so n must be odd) and each $r_i = 0$. However, any such contraction must vanish. To see this, it suffices to take $\widetilde{g}$ to be in normal form. Since one of the indices of $\widetilde{\mu}_0$ is '∞' and is contracted against one of the $\widetilde{R}$ factors, the result follows from the facts that $\widetilde{R}_{0JKL} = 0$ and $\widetilde{g}^{I\infty} = 0$ at $\rho = 0$ unless $I = 0$. $\square$

Proof of Proposition 9.1 for n even. The discussion in the n odd case is valid also for n even so long as we show that the hypotheses guarantee that the restriction of (9.2) to $\mathcal{G}$ is independent of the $O^+_{IJ}(\rho^{n/2})$ ambiguity in $\widetilde{g}$. In showing this, we may assume by Proposition 2.8 that $\widetilde{g}$ is in normal form relative to a representative metric g.

Consider the expansion of the first contraction of (9.2) into components at $\rho = 0$. By the normal form assumption, we have that $\widetilde{g}^{IJ} = 0$ at $\rho = 0$ unless $\|IJ\| = 2$. Therefore each pair of contracted indices must have strength 2 to contribute. The total number of indices being contracted is $4L + \sum r_i$. It follows that if $S_1, \ldots, S_L$ are the strengths of the factors occurring in a contributing monomial, then $\sum S_i = 4L + \sum r_i = 2L - w \leq 2L + n + 2$. As we have observed previously, Proposition 6.1 implies that for any r, a component of $\widetilde{\nabla}^r \widetilde{R}$ vanishes unless its strength is at least 4. So we may assume that $S_i \geq 4$ for each i. Thus for any i_0 we have

$$S_{i_0} + 4(L-1) \leq S_{i_0} + \sum_{i \neq i_0} S_i = \sum S_i \leq 2L + n + 2.$$

Therefore $S_{i_0} \leq n + 6 - 2L$ with strict inequality if $S_i > 4$ for some $i \neq i_0$. This implies that $S_{i_0} \leq n$ if $L \geq 3$, and that $S_{i_0} \leq n+1$ if $L = 2$ and $S_i > 4$ for some $i \neq i_0$. Proposition 6.2 guarantees in these cases that the first contraction of (9.2) is independent of the ambiguity in $\widetilde{g}$ and defines a scalar conformal invariant.

It remains to consider the case where $L = 2$ and $S_i = 4$ for some i. We may relabel the factors in (9.2) if necessary so that $S_1 = 4$. As noted above, Proposition 6.2 implies the result if $S_2 \leq n+1$, so we are reduced to consideration of the case $S_1 = 4$, $S_2 = n+2$. The only nonvanishing component with $S_1 = 4$ is $\widetilde{R}_{ijkl}$, which equals W_{ijkl} when evaluated at $\rho = 0$, $t = 1$. This must be contracted with respect to g with a component of strength $n+2$. This component of strength $n+2$ must therefore have exactly four indices between 1 and n; all other indices must be 0 or ∞. Using Proposition 6.1, we may rewrite such a component as a linear combination of components in which 0 does not occur as an index. Such a component has exactly 4 indices between 1 and n and exactly $(n-2)/2$ indices which are ∞. We investigate the dependence of such components on the $O(\rho^{n/2})$ ambiguity in the expansion of the $g_{ij}(x,\rho)$ coefficient in the ambient metric (3.14).

Consider the formula for a covariant derivative of curvature $\widetilde{R}_{IJKL,M_1\cdots M_r}$ of a general pseudo-Riemannian metric $\widetilde{g}_{IJ}$ in terms of coordinate derivatives of $\widetilde{g}_{IJ}$. Beginning with

$$\widetilde{R}_{IJKL} = \frac{1}{2}(\partial^2_{IL}\widetilde{g}_{JK} + \partial^2_{JK}\widetilde{g}_{IL} - \partial^2_{JL}\widetilde{g}_{IK} - \partial^2_{IK}\widetilde{g}_{JL}) + (\widetilde{\Gamma}^Q_{IL}\widetilde{\Gamma}_{JKQ} - \widetilde{\Gamma}^Q_{IK}\widetilde{\Gamma}_{JLQ})$$

and successively covariantly differentiating, one sees that $\widetilde{R}_{IJKL,M_1\cdots M_r} = \mathrm{I} + \mathrm{II} + \mathrm{III}$, where I is a linear combination of terms of the form $\partial^{r+2}_{P_1\cdots P_{r+2}}\widetilde{g}_{AB}$ in which $ABP_1\cdots P_{r+2}$ is a permutation of $IJKLM_1\cdots M_r$, II is a linear combination of terms of the form $\widetilde{\Gamma}^Q_{CD}\partial^{r+1}_{QP_1\cdots P_r}\widetilde{g}_{AB}$ and $\widetilde{\Gamma}^Q_{CD}\partial^{r+1}_{BP_1\cdots P_r}\widetilde{g}_{AQ}$ in which $ABCDP_1\cdots P_r$ is a permutation of $IJKLM_1\cdots M_r$, and III involves only $\partial^s\widetilde{g}$ with $s \leq r$.

Apply this observation to a curvature component $\widetilde{R}_{IJKL,M_1\cdots M_{(n-2)/2}}$ of the ambient metric, where the list of indices $IJKLM_1\cdots M_{(n-2)/2}$ is a permutation of $ijkl\underbrace{\infty\cdots\infty}_{(n-2)/2}$. When restricted to $\rho = 0$, the terms in I and III involve only $\partial^r_\rho g(x,0)$ for $r \leq n/2 - 1$. The terms in II may have a factor of $\partial^{n/2}_\rho g(x,0)$, but any such factor is multiplied by $\widetilde{\Gamma}^\infty_{ab}$, where ab are two of $ijkl$. By (3.16), we have $\widetilde{\Gamma}^\infty_{ab}|_{\rho=0} = -g_{ab}$. Consequently, such terms in II drop out when contracted against W_{ijkl}. It follows that the contraction (9.2) is independent of the $O(\rho^{n/2})$ ambiguity as desired.

The argument for the odd contractions is similar to that of the second paragraph of this proof above. We give the details for contractions of the second type. As noted previously, we may assume that $L \geq (n+2)/2$. The total number of indices being contracted is now $4L + \sum r_i + n + 2$. If S_i are the strengths of the factors in a contributing monomial in the components of the $\widetilde{\nabla}^r\widetilde{R}$, then since the strengths of the indices of $\widetilde{\mu}$ sum to $n+2$, we again have $\sum S_i = 4L + \sum r_i = 2L - w$, and now this is $\leq 2L + 2n - 2$. As in the argument above, this implies that $S_{i_0} \leq 2n + 2 - 2L \leq n$, so Proposition 6.2 implies that the contraction is independent of the ambiguity in $\widetilde{g}$. □

We remark that one can also consider the dependence on the ambiguity in $\widetilde{g}$ of the first contraction in (9.2) in case $L = 1$. Modulo contractions with $L \geq 2$, the only possibilities are $\widetilde{\Delta}^{r/2}\widetilde{S}$, where $\widetilde{S}$ denotes the scalar curvature and $\widetilde{\Delta}$ the $\widetilde{g}$-Laplacian. It is not hard to see that these are independent of the $O^+_{IJ}(\rho^{n/2})$ ambiguity in $\widetilde{g}_{IJ}$ as long as $r \leq n-2$, which corresponds to $-w \leq n$. However, the resulting conformal invariants all vanish since $\widetilde{R}_{IJ} = O^+_{IJ}(\rho^{n/2-1})$.

For an example, consider an even complete contraction (9.2) in which $r_i = 0$ for all i. Since $\widetilde{R}_{IJK0} = 0$, it follows that the resulting conformal invariant is $\operatorname{contr}(W \otimes \cdots \otimes W)$ with the same pairing of the indices, where W denotes the Weyl tensor and the contractions are now with respect to g. Of course, this is invariant with no restrictions on L when n is even.

A more interesting example is $\|\widetilde{\nabla}\widetilde{R}\|^2$. Proposition 9.1 implies that this defines a conformal invariant with $-w = 6$ in all dimensions $n \geq 3$. Expanding the contraction and evaluating the components shows that the conformal invariant is

$$\|V\|^2 + 16(W, U) + 16\|C\|^2, \tag{9.3}$$

where V_{ijklm} is given by (6.3) and

$$U_{mjkl} = C_{jkl,m} - P_m{}^i W_{ijkl}.$$

When $n = 3$ this reduces to a nonzero multiple of $\|C\|^2$. (For $n = 4$, the expression involving $B_{ij}/(n-4)$ in the formula for $\widetilde{R}_{\infty jkl,m}|_{\rho=0,\, t=1} = Y_{jklm}$ is replaced by an expression involving g''_{ij}, but this drops out as guaranteed by Proposition 9.1.)

We are interested in the question of the extent to which all scalar conformal invariants are Weyl invariants. The results presented here resolve this question for all invariants when n is odd and for invariants with $-w \leq n$ when n is even. A scalar conformal invariant is said to be *exceptional* if it is not a Weyl invariant. Theorem 9.2 shows that any nonzero odd scalar conformal invariant with $-w = n$ is exceptional. These are classified as follows.

Theorem 9.3. *If $n \not\equiv 0 \mod 4$, there are no nonzero odd scalar conformal invariants with $-w = n$.*

If $n \equiv 0 \mod 4$, the space of odd scalar conformal invariants with $-w = n$ has dimension $p(n/4)$, the number of partitions of $n/4$. Every nonzero such invariant is exceptional. A basis for this space may be taken to be the Pontrjagin invariants whose integrals give the Pontrjagin numbers of a compact oriented n-dimensional manifold.

Proof. The proof does not use conformal invariance; these are facts about the space spanned by odd contractions (9.1) with $2L + \sum r_i = n$. As noted above, we may restrict consideration to contractions with $L \geq n/2$. We must therefore have $L = n/2$ and $r_i = 0$ for each i. In particular, n must be even for such a contraction to be nonzero. Upon decomposing R into its Weyl and Ricci pieces, the fact that at most two indices of μ can be contracted against any Weyl factor implies that the Ricci curvature cannot appear in any nonzero contraction. Therefore, an odd invariant with $-w = n$ must be a linear combination of contractions of the form $\mathrm{contr}(\mu \otimes \underbrace{W \otimes \cdots \otimes W}_{n/2})$, where W is the Weyl tensor. Any such contraction is clearly conformally invariant. It is easily seen using the symmetries of W that all such contractions vanish if $n \equiv 2 \mod 4$ and that there are at most $p(n/4)$ linearly independent such contractions if $n \equiv 0 \mod 4$. The Pontrjagin invariants are $p(n/4)$ linearly independent invariants which are linear combinations of contractions of this type. It follows that the dimension of the space of such invariants is $p(n/4)$ and that the Pontrjagin invariants form a basis. □

The main result of this chapter is the following.

Theorem 9.4. *If n is odd, every scalar conformal invariant is a Weyl invariant. If n is even, every even scalar conformal invariant with $-w \leq n$ is a Weyl invariant.*

For n even, an extension of Theorems 9.3 and 9.4 covering all invariants is described in [GrH2]. A theory of even invariants of a conformal structure coupled with a conformal density is given in [Al1], [Al2] which recovers Theorem 9.4 when applied to even invariants depending only on the conformal structure.

As an illustration of the use of Theorem 9.4, consider even invariants with $-w = 4$ or 6. For $-w = 4$, the only possible even contraction (9.2) is $\|\widetilde{R}\|^2$, so all even scalar conformal invariants with $-w = 4$ are multiples of $\|W\|^2$. For even invariants with $-w = 6$, it is easily seen that modulo contractions with $L = 3$, the only possibility with $L = 2$ is $\|\widetilde{\nabla}\widetilde{R}\|^2$. It follows that all even scalar conformal invariants with $-w = 6$ (in dimension $n \neq 4$) are linear combinations of (9.3) and cubic terms in the Weyl tensor. In general, for a given weight, work is required to determine which contractions are independent, but Theorem 9.4 gives a finite list of spanning conformal invariants.

The first step in the proof of Theorem 9.4 is to reformulate scalar conformal invariants in terms of invariants of the P-action on the space of conformal normal forms. In general, by a scalar invariant of a group acting on a space is meant a scalar function on the space transforming by a character of the group. Accordingly, if $\mathcal{S}$ is a P-space, a real valued function Q on $\mathcal{S}$ is said to be an even (resp. odd) P-invariant of weight w if $Q(p.s) = \sigma_w(p)Q(s)$ (resp. $\det(p)\sigma_w(p)Q(s)$) for $s \in \mathcal{S}$, $p \in P$. Equivalently, Q defines a P-equivariant map $Q : \mathcal{S} \to \sigma_w$ (resp. $\det \otimes \sigma_w$). A function is said to be a P-invariant of $\mathcal{S}$ (or a P-invariant function on $\mathcal{S}$) if it is a sum of an even and an odd invariant of weight w for some w.

Proposition 9.5. *There is a one-to-one correspondence between scalar conformal invariants of weight w and P-invariant polynomials Q on $\mathcal{N}_c$ of weight w.*

Proof. A scalar conformal invariant $I(g)$ of weight w defines a polynomial Q on $\mathcal{N}_c$ by evaluation at $0 \in \mathbb{R}^n$. We can write $I = I_+ + I_-$, where I_+ (resp. I_-) are even (resp. odd) scalar conformal invariants of weight w. This gives $Q = Q_+ + Q_-$ for the polynomials on $\mathcal{N}_c$. The definition of the action of P on $\mathcal{N}_c$ and the invariance of I_+, I_- show immediately that Q_+ (resp. Q_-) is an even (resp. odd) invariant of $\mathcal{N}_c$ of weight w.

For the other direction, let Q be a polynomial invariant of $\mathcal{N}_c$ of weight w. If g is any metric defined near the origin in $\mathbb{R}^n$, by Proposition 8.4 and the existence of geodesic normal coordinates, we can find Ω with $\Omega(0) = 1$ and φ with $\det \varphi'(0) > 0$ such that $(\varphi^{-1})^*(\Omega^2 g) \in \mathcal{N}_c$. Define $I(g) = Q((\varphi^{-1})^*(\Omega^2 g))$. The element $(\varphi^{-1})^*(\Omega^2 g)$ of $\mathcal{N}_c$ is not uniquely determined, but Proposition 8.7 and the normalizations $\Omega(0) = 1$, $\det \varphi'(0) > 0$ imply that it is determined up to the action of an element $p \in P$ such that $a = 1$ and $\det(p) = 1$. The assumed invariance of Q therefore shows that $I(g)$ is well-defined. This well-definedness makes it clear that $I(((\varphi^{-1})^*(\Omega^2 g)) = I(g)$ for general φ, Ω such that $\det \varphi'(0) > 0$, $\Omega(0) = 1$. The transformation law $Q(p_{a^{-1}}.g) = a^w Q(g)$ for $a > 0$ and $g \in \mathcal{N}_c$ implies that $I(a^2 g) = a^w I(g)$ for $a > 0$ and general g. Therefore I satisfies the conformal transformation law $I(\Omega^2 g) = \Omega(0)^w I(g)$.

For $g \in \mathcal{N}$, the Ω and φ are uniquely determined to infinite order if we require also that $d\Omega(0) = 0$ and $\varphi'(0) = Id$, and the construction of the conformal normal form shows that the Taylor coefficients of such Ω and φ depend polynomially on $g \in \mathcal{N}$. Therefore $I(g)$ defines an $SO(h)$-invariant polynomial on $\mathcal{N}$. By the pseudo-Riemannian jet isomorphism theorem and Weyl's invariant theory, an $SO(h)$-invariant polynomial on $\mathcal{N}$ is a linear combination of complete contractions (9.1), so that $I(g)$ is a scalar invariant of metrics. Combining this with the conformal transformation law established above, we deduce that I is a scalar conformal invariant as desired.

The maps $I \to Q$ and $Q \to I$ are easily seen to be inverses of one another. □

We will use Theorem 8.10 to transfer P-invariant polynomials from $\mathcal{N}_c$ to

$\widetilde{\mathcal{R}}$. The next lemma will assure when n is even that we are in the range in which Theorem 8.10 applies.

Lemma 9.6. *Let $I(g)$ be an even scalar conformal invariant of weight w. The associated polynomial $Q(g)$ on $\mathcal{N}_c$ determined by Proposition 9.5 can be written as the restriction to $\mathcal{N}_c$ of a polynomial in derivatives of g of order $\leq -w-2$.*

Proof. The polynomial $Q(g)$ determined by Proposition 9.5 agrees with the restriction to $\mathcal{N}_c$ of a linear combination of even complete contractions (9.1) with $2L + \sum r_i = -w$. Thus for each i_0 we have $r_{i_0} \leq \sum r_i = -w - 2L$. If $L \geq 2$, we obtain $r_{i_0} \leq -w - 4$, so the contraction involves at most $-w-2$ derivatives of g as claimed. Modulo contractions with $L \geq 2$, the only possibility with $L = 1$ is $\Delta^{(-w-2)/2}S$, where S denotes the scalar curvature. But contracting $\mathrm{Sym}(\nabla^r \mathrm{Ric})(g)(0) = 0$ and using the Bianchi and Ricci identities shows that $\Delta^{(-w-2)/2}S$ agrees on $\mathcal{N}_c$ with a linear combination of complete contractions with $L \geq 2$. □

Let $\widetilde{\varepsilon}$ denote the usual volume form on $\mathbb{R}^{n+2}$ and $\widetilde{\varepsilon}_0 = e_0 \lrcorner \widetilde{\varepsilon}$. Complete contractions with respect to $\widetilde{h}$ of the form

$$\begin{gathered}\mathrm{contr}(\widetilde{R}^{(r_1)} \otimes \cdots \otimes \widetilde{R}^{(r_L)})\\ \mathrm{contr}(\widetilde{\varepsilon} \otimes \widetilde{R}^{(r_1)} \otimes \cdots \otimes \widetilde{R}^{(r_L)})\\ \mathrm{contr}(\widetilde{\varepsilon}_0 \otimes \widetilde{R}^{(r_1)} \otimes \cdots \otimes \widetilde{R}^{(r_L)})\end{gathered} \tag{9.4}$$

define P-invariant polynomials on $\widetilde{\mathcal{T}}$ with $-w = 2L + \sum r_i$ for the first two and $-w = 2L + \sum r_i - 1$ for the third. By a Weyl invariant of weight w of $\widetilde{\mathcal{R}}$ (resp. $T\widetilde{\mathcal{R}}$), we shall mean the restriction to $\widetilde{\mathcal{R}}$ (resp. $T\widetilde{\mathcal{R}}$) of a linear combination of contractions (9.4) of weight w.

Lemma 9.7. *When restricted to $\widetilde{\mathcal{R}}$, the first contraction in* (9.4) *defines a P-invariant polynomial on $\widetilde{\mathcal{R}}^{-w-4}$, i.e., it factors through the projection $\widetilde{\mathcal{R}} \to \widetilde{\mathcal{R}}^{-w-4}$. The same statement also holds with $\widetilde{\mathcal{R}}$ replaced everywhere by $T\widetilde{\mathcal{R}}$.*

Proof. The argument of the second paragraph of the proof of Proposition 9.1 shows that when expanded, the first contraction in (9.4) involves only components $\widetilde{R}_{IJKL,\mathcal{M}}$ with $\|IJKL\mathcal{M}\| \leq 4 - 2L - w$. The defining relations of $\widetilde{\mathcal{R}}$ and $T\widetilde{\mathcal{R}}$ show that when restricted to either of these spaces, any such contraction with $L = 1$ can be rewritten as a linear combination of contractions with $L \geq 2$. This gives the bound $\|IJKL\mathcal{M}\| \leq -w$, which is the desired statement. □

The main result we will use characterizing invariants of $\widetilde{\mathcal{R}}$ is the following.

Theorem 9.8. *If n is odd, every P-invariant polynomial on $\widetilde{\mathcal{R}}$ is a Weyl invariant. If n is even, every even P-invariant polynomial on $\widetilde{\mathcal{R}}$ is a Weyl invariant.*

Proof of Theorem 9.4 using Theorem 9.8. For n odd, Proposition 9.5 and Theorem 8.10 immediately reduce Theorem 9.4 to Theorem 9.8. For n even, an even scalar conformal invariant $I(g)$ with $-w \leq n$ determines by Proposition 9.5 and Lemma 9.6 an even P-invariant polynomial of weight w on $\mathcal{N}_c^{n-2}$. Theorem 8.10 shows that this defines an even P-invariant polynomial of weight w on $\widetilde{\mathcal{R}}^{n-4}$. Theorem 9.8 asserts that as a function on $\widetilde{\mathcal{R}}$, this polynomial agrees with a linear combination of complete contractions (9.4) of weight w, and Lemma 9.7 shows that each of these can be regarded as a P-invariant of weight w of $\widetilde{\mathcal{R}}^{n-4}$. Reversing the steps, it follows that $I(g)$ has the desired form. □

Theorem 9.8 is proved by reduction to the following, which is one of the main results of [BEGr].

Theorem 9.9. *[BEGr] If n is odd, every P-invariant polynomial on $T\widetilde{\mathcal{R}}$ is a Weyl invariant. If n is even, every even P-invariant polynomial on $T\widetilde{\mathcal{R}}$ is a Weyl invariant.*

Proof of Theorem 9.8 using Theorem 9.9. We first extend the notion of weight to polynomials which are not P-invariant. A polynomial on $\widetilde{\mathcal{T}}$ is a polynomial in the variables $(\widetilde{R}_{IJKL,\mathcal{M}})_{|\mathcal{M}|\geq 0}$, subject to the linear relations defining $\widetilde{\mathcal{T}}$. The degree of such a polynomial will refer to its degree in the variables $(\widetilde{R}_{IJKL,\mathcal{M}})$. A polynomial Q on $\widetilde{\mathcal{T}}$ will be said to have weight w if it satisfies $Q(p_a.(\widetilde{R})) = \sigma_w(p_a)Q((\widetilde{R}))$ for $a > 0$. The linear polynomial $\widetilde{R}_{IJKL,\mathcal{M}}$ has $-w = \|IJKL\mathcal{M}\| - 2$. Since $\widetilde{R}_{IJKL,\mathcal{M}}$ vanishes unless $\|IJKL\mathcal{M}\| \geq 4$, one sees that a monomial in the variables $(\widetilde{R}_{IJKL,\mathcal{M}})$ must vanish if its weight w and degree d satisfy $d > -w/2$. Observe that each of the relations (1)–(6) in Definition 8.8 defining $\widetilde{\mathcal{R}}$ states the vanishing of a polynomial of a specific weight, i.e., $\widetilde{\mathcal{R}}$ is defined by polynomials homogeneous with respect to the action of the p_a.

Let Q be a P-invariant polynomial of weight w on $\widetilde{\mathcal{R}}$, assumed to be even if n is even. We will show by induction that for any $d \geq 0$, Q can be written in the form $Q = Q_d + E_d$ on $\widetilde{\mathcal{R}}$, where Q_d is a linear combination of complete contractions (9.4) of weight w and E_d is a polynomial on $\widetilde{\mathcal{T}}$ of weight w which is the sum of monomials of degree $\geq d$. The observation above implies that $E_d = 0$ for $d > -w/2$, which gives the desired conclusion.

The induction statement is clear for $d = 0$. Suppose it holds for d, so that $Q = Q_d + E_d$ on $\widetilde{\mathcal{R}}$. Write $E_d = E'_d + E''_d$, where E'_d and E''_d are polynomials on $\widetilde{\mathcal{T}}$ of weight w, E'_d is homogeneous of degree d, and E''_d is a linear combination of monomials of degree $> d$. Since E_d is P-invariant when restricted to $\widetilde{\mathcal{R}}$, it follows easily using the first part of Theorem 8.10 that E'_d is P-invariant of weight w when restricted to $T\widetilde{\mathcal{R}}$. So by Theorem 9.9, there is a linear combination of complete contractions Q'_d of weight w, which we may take to be homogeneous of degree d as polynomials on $\widetilde{\mathcal{T}}$, such that $E'_d = Q'_d$ on $T\widetilde{\mathcal{R}}$. Let P_i, $i = 1, 2, \dots$ be an enumeration of the polynomials on $\widetilde{\mathcal{T}}$ whose vanishing defines $\widetilde{\mathcal{R}}$, and let p_i denote the linear part of P_i, so that $T\widetilde{\mathcal{R}}$ is defined by the vanishing of the p_i. We conclude that we can write $E'_d - Q'_d = \sum U_i p_i$ as polynomials on $\widetilde{\mathcal{T}}$, where the U_i are homogeneous polynomials of degree $d-1$ and the sum is finite. When restricted to $\widetilde{\mathcal{R}}$, we have $E'_d - Q'_d = \sum U_i(p_i - P_i)$. So if we set $Q_{d+1} = Q_d + Q'_d$ and $E_{d+1} = E''_d + \sum U_i(p_i - P_i)$, then we have $Q = Q_{d+1} + E_{d+1}$ on $\widetilde{\mathcal{R}}$, where Q_{d+1} is a linear combination of complete contractions of weight w and E_{d+1} is a polynomial containing only monomials of degree at least $d+1$. This completes the induction step. □

We remark that Theorem 9.9 also holds for odd invariants when $n \equiv 2$ mod 4, by combining results in [BEGr] and [BaiG]. The argument given above thus shows that Theorem 9.8 holds also for odd invariants when $n \equiv 2$ mod 4. Of course, the jet isomorphism theorem of Chapter 8 does not apply to higher order jets in even dimensions, and $\widetilde{\mathcal{R}}$ is no longer the correct space to study to understand conformal invariants.

Bibliography

[Al1] S. Alexakis, *On conformally invariant differential operators in odd dimensions*, Proc. Natl. Acad. Sci. USA **100** (2003), 4409-4410.

[Al2] S. Alexakis, *On conformally invariant differential operators*, arXiv:math/0608771.

[Ar] S. Armstrong, *Definite signature conformal holonomy: a complete classification*, J. Geom. Phys. **57** (2007), 2024-2048, arXiv:math/0503388.

[AL] S. Armstrong and T. Leistner, *Ambient connections realising conformal tractor holonomy*, Monatsh. Math. **152** (2007), 265-282, arXiv:math/0606410.

[AW] R. Askey and J. Wilson, *Some basic hypergeometric orthogonal polynomials that generalize Jacobi polynomials*, Mem. A. M. S. **54** no. 319 (1985).

[ABP] M. Atiyah, R. Bott and V. K. Patodi, *On the heat equation and the index theorem*, Invent. Math. **19** (1973), 279-330. Errata: **28** (1975), 277–280.

[BEGo] T. N. Bailey, M. G. Eastwood and A. R. Gover, *Thomas's structure bundle for conformal, projective and related structures*, Rocky Mountain J. Math. **24** (1994), 1191-1217.

[BEGr] T. N. Bailey, M. G. Eastwood and C. R. Graham, *Invariant theory for conformal and CR geometry*, Ann. Math. **139** (1994), 491-552.

[BaiG] T. N. Bailey and A. R. Gover, *Exceptional invariants in the parabolic invariant theory of conformal geometry*, Proc. A. M. S. **123** (1995), 2535-2543.

[BaoG] M. S. Baouendi and C. Goulaouic, *Singular nonlinear Cauchy problems*, J. Diff. Eq. **22** (1976), 268-291.

[Be] A. Besse, *Einstein Manifolds*, Springer, 2002.

[Br] T. Branson, *Sharp inequalities, the functional determinant, and the complementary series*, Trans. A. M. S. **347** (1995), 3671-3742.

[BEGW] T. Branson, M. Eastwood, A. R. Gover and M. Wang, Organizers, *Conformal Structure in Geometry, Analysis and Physics*, AIM Workshop Summary, Aug. 2003, http://www.aimath.org/pastworkshops/confstructrep.pdf.

[BrG] T. Branson and A. R. Gover, *Conformally invariant operators, differential forms, cohomology and a generalisation of Q-curvature*, Comm. P. D. E. **30** (2005), 1611-1669, arXiv:math/0309085.

[ČG] A. Čap and A. R. Gover, *Standard tractors and the conformal ambient metric construction*, Ann. Global Anal. Geom. **24** (2003), 231-259.

[E1] C. Epstein, *An asymptotic volume formula for convex cocompact hyperbolic manifolds*, Appendix A in *The divisor of Selberg's zeta function for Kleinian groups*, S. J. Patterson and P. A. Perry, Duke Math. J. **106** (2001), 321-390.

[E2] D. Epstein, *Natural tensors on Riemannian manifolds*, J. Diff. Geom. **10** (1975), 631-645.

[F] C. Fefferman, *Parabolic invariant theory in complex analysis*, Adv. Math. **31** (1979), 131-262.

[FG] C. Fefferman and C. R. Graham, *Conformal invariants,* in *The Mathematical Heritage of Élie Cartan (Lyon, 1984)*, Astérisque, 1985, Numero Hors Serie, 95-116.

[FH] C. Fefferman and K. Hirachi, *Ambient metric construction of Q-curvature in conformal and CR geometries*, Math. Res. Lett. **10** (2003), 819-832.

[Go1] A. R. Gover, *Invariant theory and calculus for conformal geometries*, Adv. Math. **163** (2001), 206-257.

[Go2] A. R. Gover, *Laplacian operators and Q-curvature on conformally Einstein manifolds*, Math. Ann. **336** (2006), 311-334, arXiv:math/0506037.

[GoH] A. R. Gover and K. Hirachi, *Conformally invariant powers of the Laplacian–a complete non-existence theorem*, J. A. M. S. **17** (2004), 389-405, arXiv:math/0304082.

[GoL] A. R. Gover and F. Leitner, *A sub-product construction of Poincaré-Einstein metrics*, Internat. J. Math. **20** (2009), 1263-1287, arXiv:math/0608044.

[GP1] A. R. Gover and L. J. Peterson, *Conformally invariant powers of the Laplacian, Q-curvature, and tractor calculus*, Comm. Math. Phys. **235** (2003), 339-378, arXiv:math-ph/0201030.

[GP2] A. R. Gover and L. J. Peterson, *The ambient obstruction tensor and the conformal deformation complex*, Pacific J. Math. **226** (2006), 309-351, arXiv:math/0408229.

[Gr1] C. R. Graham, *Conformally invariant powers of the Laplacian, II: Nonexistence*, J. London Math. Soc. **46** (1992), 566-576.

[Gr2] C. R. Graham, *Dirichlet-to-Neumann map for Poincaré-Einstein metrics*, Oberwolfach Rep. **39** (2005), 2200-2203.

[Gr3] C. R. Graham, *Jet isomorphism for conformal geometry*, Arch. Math. (Brno): Proc. 27th Czech Winter School on Geometry and Physics (Srní, Jan. 2007) **43** (2007), 389-415, arXiv:0710.1671.

[GrH1] C. R. Graham and K. Hirachi, *The ambient obstruction tensor and Q-curvature*, in *AdS/CFT Correspondence: Einstein Metrics and their Conformal Boundaries*, IRMA Lectures in Mathematics and Theoretical Physics **8** (2005), 59-71, arXiv:math/0405068.

[GrH2] C. R. Graham and K. Hirachi, *Inhomogeneous ambient metrics*, IMA Vol. Math. Appl. **144**: Symmetries and Overdetermined Systems of Partial Differential Equations (2008), 403-420, arXiv:math/0611931.

[GJMS] C. R. Graham, R. Jenne, L. J. Mason and G. A. J. Sparling, *Conformally invariant powers of the Laplacian, I: Existence*, J. London Math. Soc. **46** (1992), 557-565.

[GJ] C. R. Graham and A. Juhl, *Holographic formula for Q-curvature*, Adv. Math. **216** (2007), 841-853, arXiv:0704.1673.

[GrL] C. R. Graham and J. M. Lee, *Einstein metrics with prescribed conformal infinity on the ball*, Adv. Math. **87** (1991), 186-225.

[HS] J. Haantjes and A. J. Schouten, *Beiträge zur allgemeinen (gekrümmten) konformen differentialgeometrie, I, II*, Math. Ann. **112** (1936), 594-629, **113** (1937), 568-583.

[HSS] S. de Haro, K. Skenderis and S. N. Solodukhin, *Holographic reconstruction of spacetime and renormalization in the AdS/CFT correspondence*, Comm. Math. Phys. **217** (2001), 594-622, arXiv:hep-th/0002230.

[Hi] K. Hirachi, *Construction of boundary invariants and the logarithmic singularity of the Bergman kernel*, Ann. Math. **151** (2000), 151-191.

[Hö] L. Hörmander, *The Analysis of Linear Partial Differential Operators III*, Springer, 1985.

[J] A. Juhl, *Families of Conformally Covariant Operators, Q-Curvature and Holography*, Birkhäuser, 2009.

[KM] S. Karlin and J. L. McGregor, *The Hahn polynomials, formulas and an application*, Scripta Math. **26** (1961), 33-46.

[K] S. Kichenassamy, *On a conjecture of Fefferman and Graham*, Adv. Math. **184** (2004), 268-288.

[L] S. Lang, *Fundamentals of Differential Geometry*, Springer, 1999.

[LeB] C. LeBrun, *$\mathcal{H}$-space with a cosmological constant*, Proc. Roy. Soc. London A **380** (1982), 171-185.

[LP] J. M. Lee and T. H. Parker, *The Yamabe problem*, Bull. A. M. S. **17** (1987), 37-91.

[Leis] T. Leistner, *Conformal holonomy of C-spaces, Ricci-flat, and Lorentzian manifolds*, Diff. Geom. Appl. **24** (2006), 458-478, arXiv:math/0501239.

[Leit] F. Leitner, *Normal conformal Killing forms*, arXiv:math/0406316.

[M] R. R. Mazzeo, *The Hodge cohomology of a conformally compact metric*, J. Diff. Geom. **28** (1988), 309-339.

[NSU] A. F. Nikiforov, S. K. Suslov and V. B. Uvarov, *Classical Orthogonal Polynomials of a Discrete Variable*, Springer, 1991.

[N] P. Nurowski, *Conformal structures with explicit ambient metrics and conformal G_2 holonomy*, IMA Vol. Math. Appl. **144**: Symmetries and Overdetermined Systems of Partial Differential Equations (2008), 515-526, arXiv:math/0701891.

[R] A. Rendall, *Asymptotics of solutions of the Einstein equations with positive cosmological constant*, Ann. Henri Poincaré **5** (2004), 1041-1064, arXiv:gr-qc/0312020.

[SS] K. Skenderis and S. N. Solodukin, *Quantum effective action from the AdS/CFT correspondence*, Phys. Lett. **B472** (2000), 316-322, arXiv:hep-th/9910023.

[W] H. Weyl, *The Classical Groups*, Princeton University Press, 1939.

Index